HUMAN RESOURCES AND CREW RESOURCE MAN[illegible]

Human Resources and Crew Resource Management

Marine Navigation and Safety of Sea Transportation

Editors

Adam Weintrit & Tomasz Neumann

Gdynia Maritime University, Gdynia, Poland

CRC Press is an imprint of the
Taylor & Francis Group, an **informa** business
A BALKEMA BOOK

CRC Press/Balkema is an imprint of the Taylor & Francis Group, an informa business

Printed and bound in Great Britain by Antony Rowe Ltd (A CPI-group Company), Chippenham, Wiltshire

Published by: CRC Press/Balkema
P.O. Box 447, 2300 AK Leiden, The Netherlands
e-mail: Pub.NL@taylorandfrancis.com
www.crcpress.com – www.taylorandfrancis.co.uk – www.balkema.nl

ISBN: 978-0-415-69115-4 (Pbk)
ISBN: 978-0-203-15729-9 (eBook)

List of reviewers

Prof. Yasuo **Arai**, President of Japan Institute of Navigation, Japan,
Prof. Andrzej **Banachowicz**, West Pomeranian University of Technology, Szczecin, Poland,
Prof. Eugen **Barsan**, Master Mariner, Constanta Maritime University, Romania,
Sr. Jesus **Carbajosa Menendez**, President of Spanish Institute of Navigation, Spain,
Prof. Doina **Carp**, Constanta Maritime University, Romania,
Prof. Shyy Woei **Chang**, National Kaohsiung Marine University, Taiwan,
Prof. Adam **Charchalis**, Gdynia Maritime University, Poland,
Prof. Krzysztof **Chwesiuk**, Maritime University of Szczecin, Poland,
Prof. Jerzy **Czajkowski**, Gdynia Maritime University, Poland,
Prof. Krzysztof **Czaplewski**, Polish Naval Academy, Gdynia, Poland,
Prof. German **de Melo Rodrigues**, Technical University of Catalonia, Barcelona, Spain,
Prof. Daniel **Duda**, Master Mariner, President of Polish Nautological Society, Naval University of Gdynia, Polish Nautological Society, Poland,
RAdm. Dr. Czesław **Dyrcz**, Rector of Polish Naval Academy, Naval Academy, Gdynia, Poland,
Prof. Wiliam **Eisenhardt**, President of the California Maritime Academy, Vallejo, USA,
Prof. Wlodzimierz **Filipowicz**, Master Mariner, Gdynia Maritime University, Poland,
Prof. Avtandil **Gegenava**, Batumi Maritime Academy, Georgia,
Prof. Stanislaw **Gorski**, Master Mariner, Gdynia Maritime University, Poland,
Prof. Andrzej **Grzelakowski**, Gdynia Maritime University, Poland,
Prof. Jerzy **Hajduk**, Master Mariner, Maritime University of Szczecin, Poland,
Prof. Michal **Holec**, Gdynia Maritime University, Poland,
Prof. Bogdan **Jaremin**, Interdepartmental Institute of Maritime and Tropical Medicine in Gdynia,,
Prof. Yongxing **Jin**, Shanghai Maritime University, China,
Prof. Mirosław **Jurdzinski**, Master Mariner, FNI, Gdynia Maritime University, Poland,
Prof. Serdjo **Kos**, FRIN, University of Rijeka, Croatia,
Prof. Andrzej **Krolikowski**, Master Mariner, Maritime Office in Gdynia, Poland,
Prof. Pentti Kujala, **Helsinki** University of Technology, Helsinki, Finland,
Prof. Bogumil **Laczynski**, Master Mariner, Gdynia Maritime University, Poland,
Prof. Andrzej S. **Lenart**, Gdynia Maritime University, Poland,
Prof. Vladimir **Loginovsky**, Admiral Makarov State Maritime Academy, St. Petersburg, Russia,
Prof. Evgeniy **Lushnikov**, Maritime University of Szczecin, Poland,
Prof. Melchor M. **Magramo**, John B. Lacson Foundation Maritime University, Iloilo City, Philippines,
Prof. Boyan **Mednikarov**, Nikola Y. Vaptsarov Naval Academy,Varna, Bulgaria,
Prof. Janusz **Mindykowski**, Gdynia Maritime University, Poland,
Prof. Mykhaylo V. **Miyusov**, Rector of Odesa National Maritime Academy, Odesa, Ukraine,
Prof. Proshanto **Mukherjee**, FNI, AFRIN, World Maritime University, Malmoe, Sweden,
Prof. Takeshi **Nakazawa**, World Maritime University, Malmoe, Sweden,
Prof. Washington Yotto **Ochieng**, Imperial College London, United Kingdom,
Prof. Wiesław **Ostachowicz**, Gdynia Maritime University, Poland,
Prof. Marcin **Plinski**, University of Gdansk, Poland,
Prof. Boris **Pritchard**, University of Rijeka, Croatia,
Prof. Piotr **Przybylowski**, Gdynia Maritime University, Poland,
Prof. Wladysław **Rymarz**, Master Mariner, Gdynia Maritime University, Poland,
Prof. Aydin **Salci**, Istanbul Technical University, Maritime Faculty, ITUMF, Istanbul, Turkey,
Prof. Viktoras **Sencila**, Lithuanian Maritime Academy, Klaipeda, Lithuania,
Prof. Roman **Smierzchalski**, Gdańsk University of Technology, Poland,
Prof. Henryk **Sniegocki**, Master Mariner, MNI, Gdynia Maritime University, Poland,
Cmdr. Bengt **Stahl**, Nordic Institute of Navigation, Sweden,
Prof. Mykola **Tsymbal**, Odessa National Maritime Academy, Ukraine,
Prof. George Yesu Vedha **Victor**, International Seaport Dredging Limited, Chennai, India,
Prof. Vladimir **Volkogon**, Rector of Baltic Fishing Fleet State Academy, Kaliningrad, Russia,
Mr. Philip **Wake**, FNI, Chief Executive The Nautical Institute, London, UK,
Prof. Ryszard **Wawruch**, Master Mariner, Gdynia Maritime University, Poland,
Prof. Jia-Jang **Wu**, National Kaohsiung Marine University, Kaohsiung, Taiwan (ROC),
Prof. Homayoun **Yousefi**, MNI, Chabahar Maritime University, Iran,
Prof. Wu **Zhaolin**, Dalian Maritime University, China

Contents

Human Resources and Crew Resource Management. Introduction

A. Weintrit & T. Neumann
Gdynia Maritime University, Gdynia, Poland

PREFACE

The contents of the book are partitioned into seven parts: crew resource management (covering the chapters 1 through 4), human factors (covering the chapters 5 through 6), STCW Convention (covering the chapters 7 through 11), maritime education and training (covering the chapters 12 through 21), piracy problem (covering the chapters 22 through 25), health problems (covering the chapter 26 only), maritime ecology (covering the chapters 27 through 29). Certainly, this subject may be seen from different perspectives.

The first part deals with crew resource management. The contents of the first part are partitioned into four chapters: Crew resource management: the role of human factors and bridge resource management in reducing maritime casualties, Women seafarers: solution to shortage of competent officers?, The manning companies in the Philippines amidst the global financial crisis, and Academe and industry collaboration: key to more competent officers?

The second part deals with human factors. The contents of the second part are partitioned into two chapters: Factors of human resources competitiveness in maritime transport, and Human factors as causes for shipboard oil pollution violations.

The third part deals with STCW Convention. The contents of the third part are partitioned into five chapters: Needs and importance of master studies for navigators in XXI century and connectivity to STCW 78/95; Implementation of the 1995 STCW convention in Constanta Maritime University, Implementation of STCW Convention at the Serbian Military Academy, Electrical, electronic and control engineering – new mandatory standards of competence for engineer officers, regarding Provisions of the Manila Amendments to the STCW Code, and Assessment components influencing effectiveness of studies: marine engineering students opinion.

The fourth part deals with maritime education and training. The contents of the fourth part are partitioned into ten chapters: Improving MET quality: relationship between motives of choosing maritime professions and students' approaches to learning, Evaluation of educational software for marine training with the aid of neuroscience methods and tools, Methodological approaches to the design of business games and definition of the marine specialist training content, Teaching cross cultural competence in maritime schools, Considerations on maritime watch keeping officers' vocational training, Simulation training for replenishment at sea (RAS) operations: addressing the unique problems of 'close-alongside' and 'in-line' support for multi-streamer seismic survey vessels underway, Teaching of ROR or learning of ROR, Safety and security trainer SST7 – a new way to prepare crews managing emergency situations, The MarEng Plus project and the new applications, and Methods of maritime-related word stock research in the practical work of a maritime English teacher

The fifth part deals with piracy problem. The contents of the fifth part are partitioned into four chapters: Somali Piracy: relation between crew nationality and a vessel's vulnerability to seajacking, Influence of pirates' activities on maritime transport in the Gulf of Aden region, Preventiv[illegible]ons and safety measures directed against pir[illegible] Gulf of Aden region, and Technological [illegible] forts to reduce piracy.

The sixth part deals with [illegible] covers one chapter only: Sy[illegible]nd control of communicable [illegible]

The seventh part d[illegible]cology. The contents of the [illegible]oned into four chapters: Stu[illegible]n China international ship[illegible]ne to extreme flood and ero[illegible]y climate changes: study case[illegible]ver bar navigation in Caraguatatuba [illegible]ate, Brazil), Ecological risk from ball[illegible]rs for the harbour in Świnoujście, and A S[illegible]y Assurance Assessment

Model for an Liquefied Natural Gas (LNG) Tanker Fleet.

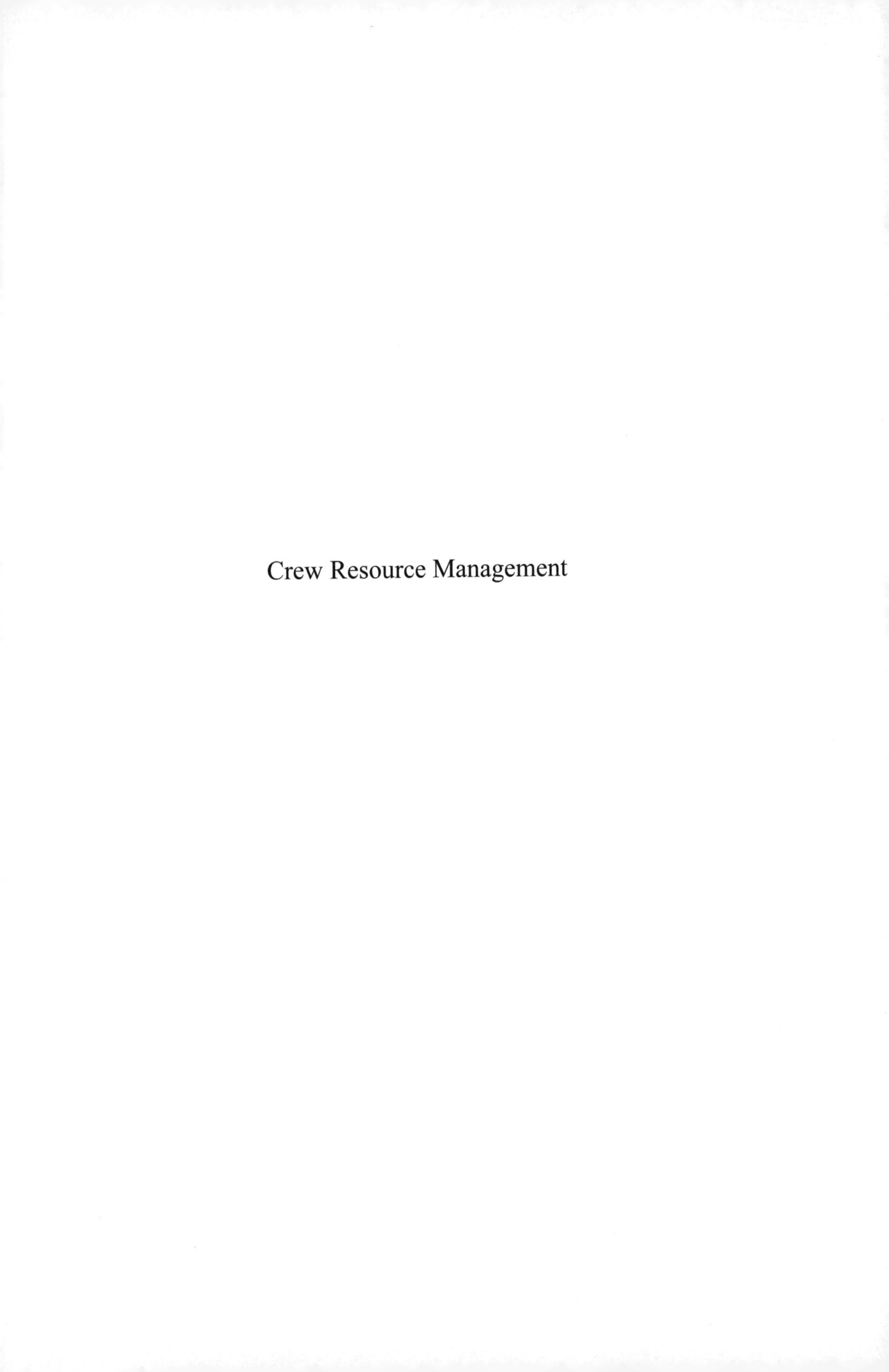

Crew Resource Management

1. Crew Resource Management: The Role of Human Factors and Bridge Resource Management in Reducing Maritime Casualties

H. Yousefi
Khoramshahar Maritime University, Khoramshahar, Iran

R. Seyedjavadin
Business Management College, Tehran University, Tehran, Iran

ABSTRACT: This paper presents the Crew Resource Management which has now been in the existence for more than two decades as a foundation of maritime transport in order to improve the operational efficiency of shipping. The impact of human errors on collisions and grounding of ocean going vessels have been taken place due to the human or team errors which need to be analyzed by various maritime casualties in depth. The first section of this article is devoted to investigate the role of Human Resource Management, Crew Resource Management and Maritime Crew Resource Management; it is because of minimizing ship accidents at sea. The next part of this paper is designated to evaluate the Bridge Resource Management, Bridge Team Management and Human factors in depth. It should be noted that the necessary techniques in bridge team management should be clarified based on the consideration of the issues that why bridge team management is arranged. The next segment of this paper is dedicated to consider the ways of minimizing ship accidents by offering optimum training methods for the future seafarers. The last part of this paper is designated to assess the qualification of maritime lecturers based on STCW95 Convention and the MARCON project for teaching the Bridge Resource Management.

1 HUMAN RESOURCE MANAGEMENT

Human Resource Management is an innovative view of the workplace management which is established as useful method for analysing the strategic approach to the organizational management. This method of management system has been defined from different sources in a similar manner, for instance, While Miler (1987) who stated that ".......those decisions and actions which concern the management of employees at all levels in the business and which are related to the implementation of strategies directed towards creating and sustaining competitive advantage". Miler emphasised on actions related to the management employee in order to maintain competitive advantage. It should be noted that after about a decade the definition of Human Resource Management (HRM) is highlighted on its responsibility for staffing people into the organization. In this regard Cherrington (1995) expressed that "Human resource management is responsible for how people are treated in organizations. It is responsible for bringing people into the organization, helping them perform their work, compensating them for their labours, and solving problems that arise".

The initiation of the advanced theory of management caused that the term of describing the function of workforce to be changed from "Personnel" to "Industrial relations" to "employee relations" and finally to "Human Resource". Although, all the above mentioned terms are used nevertheless, the precise and useful term is Human resource Management. Some people believed that HRM is a part of HRD (Human Resource Development) which includes the great range of activities in order to develop personnel of organizations; in fact, the objective of HRM is to increase the productivity of an organization by improving the effectiveness of its employees. In the last three decades, many changes have been applied to the HRM function and HRD profession; in the past, large organizations looked to the "Personnel Department," nevertheless recently, organizations consider the "HR Department" as playing an important role in staffing, training and helping to manage people. The link between human resource management and the strategic goals of an organization has been investigated by Miller (1989) as "The goal of human resource management is to help an organization to meet strategic goals by attracting, and maintaining employees and also to manage them effectively". After World War II, more attentions needed to be applied to the labours due to the shortage of skilled workers of companies, therefore the general concentrate of HRM modified from concentration on labour efficiency and skills to employee satisfaction. Then, as consequences of the Act of

1960 and Act of 1970, companies began putting more accents on HRM in order to avoid of violating this legislations. In 1980, the HRM grows up rapidly due to the several reasons such as the organizations required skills of HRM professionals in order to adapt the organization structure with a new generation of labour attitudes and behaviours, improving educational levels, growth of offering services, white colour job and more women as workforce, etc. Marine Accident Investigation Branch (1999) stated that four factors or four "Cs" such as commitment, competence, cost-effectiveness, and congruence should be used in order to evaluate whether the HRM programme is succeeded or not? The author of this paper expand the above assessment to the following five factors by adding Customer as five item in order to determine whether the HRM programme is succeeded or not. It is because; the services of Human resource in an organization are provided for the customer satisfaction, therefore we should pay careful attention to attract our customers by improving the efficiency of the organization. In 1990, the new technologies such as IT, satellite communications, computers and networking systems, fax machines, and other new devices have forced to change the field of HRM. The second important change influencing HRM was related to the recent organizational structures, which emerged during the 1980s as a result of the operational expansions of many companies or diversification of their products and services that continued through the 1990s. The third or last factor of forcing to change the HRM field is market globalization through world trade which was enhancing competition abroad; in order to compensate with the international competition, companies should consider their HRM professionals for improving the quality of products, productivity, and innovation of the organization.

2 CREW RESOURCE MANAGEMENT

Crew Resource Management has come to light for more than two decades, nevertheless a misunderstanding still exists within aviation and shipping industry for what the term entails. Crew Resource Management (CRM) is a reliable management system which makes the best use of all available means such as equipment, people in order to improve safety and increase operational activities within the shipping industry. CRM includes a broad variety of knowledge and skills encompassing cognitive skills and interpersonal skills. ....very often extremely traditional education forms and context are very demotivating of our young generations, Stephen J.Cross (2010). The cognitive skills are regarded as the psychological procedures used for acquiring situational awareness, for solving problems, and for taking decision and the interpersonal skills consist of communications and teamwork. Crew Resource Management is a management system which creates the best possible use of all available means such as equipment, communication, process and people in order to improve safety of ship operations.

For the purpose of increasing the effectiveness of crew members on board ships, a special training course of the technical knowledge not only is essential for crew members; but also they require to implement the requirement of CRM which are developing and understanding of the cognitive and interpersonal skills.

Training of CRM skills are related to the theory and practice of group behaviours which are properly developed through a method called experiential learning. It should be noted that CRM skills have been taught individually as technical knowledge and skills, nevertheless the considerable area of overlap between the above two methods proposes that the training would be more effective if it was included from the basic of the Maritime Crew training method. Dominic Cardozo (1993) defined that Crew Resource Management is an active and interactive process including all crew members, which helps to detect, communicate and avoid or handle significant threats to an operation, action or task by developing and applying efficient countermeasures to minimize the safety risk. Hackman (1986) describes the crew as being frontline operators and safety to be a line function by which the entire company culture and company structure is defined. For the purpose of assessing CRM in the maritime industries, it is necessary to have a look at the growth of CRM in aviation. In 1980, the first CRM training course was implemented by the United Airlines which was about the evaluation of crew members of their own and their team members' performance. Nowadays, CRM is an essential training course not only for all airlines and aviation schools, but also is mandatory for maritime colleges in order to reduce human errors and to avoid any incident, accident or /and collision. Since the main reasons of the majority of ship accidents can be found in human errors, therefore, CRM plays a vital role in reducing their negative result. CRM training course is necessary for operators in every work field especially for ship's crew based on the latest revision of the STCW in June 2010. In a sense, the overall integration of each responsibility on board ship as a complicated teamwork is called designing CRM training and application for the maritime industry. It means that deck and engine officers, OS, AB, cooks and stewards with different nationalities working closely together as a team. As a result of the technological improvement in shipping, the implementation of CRM strategies has been taken more than three decades which is not offer a completely agreeable result; therefore the old schemes of designing CRM training need to be updated. It should be noted that since the situation in aviation is completely different with shipping business, therefore it is not possible to copy or transfer CRM from aviation to maritime field. It is because a ship is a work place which cannot be left by the crew for months; navigators are working on board ship for 24 hours on seven days a week in order to keep the sea watch as duty officers who are not allow to sleep more than six hours.

3 BRIDGE RESOURCE MANAGEMENT

It is a training course in order to manage ship bridge activities by Master, Pilot, Watchman, Wheelman and officer on watch. Many collisions occurred as a result of misunderstanding of the parties doing different activities on the bridge for instance pilots with watch keepers. It is possible to imitate the expected situations by using ship handling simulators in order to improve the skill and communication between the responsible persons on the bridge. For instance, in 1974, Large Crude Carrier (VLCC) called "Metulla" grounded in the Magellan Straits with two pilots and watch keepers who were present on the bridge; it means that bridge teams were not working efficiently in order to support each other. In fact, there are two different comments on using simulator as training course, first Gyles and Salmon (1978) who stated that "Simulator-based training courses were introduced primarily to train the skills of passage planning and the importance of the Master/Pilot relationship". This training initiative developed into the Bridge Team Management (BTM) courses that are conducted today on many simulators world-wide and contain many of the elements to be found in CRM courses in other industries. The role of simulator as a tool for training CRM has been stated by Barnet (2060) as follows "Bridge Resource Management (BRM) courses are a more recent initiative, adapted directly from the aviation model, and are not always based on the use of simulators". Nowadays, Maritime Universities offer a training course called Bridge Team Management (BTM) or Bridge Resource Management (BRM) which is about the discussion of ship handling and navigational skills, unfortunately not emphasis on human factor. Micheal Barnnet (2060) quoted from Flin et al regarding the significance of the bridge resource management course in the following paragraph as follows "Within other safety critical industries, and the military, the training and assessment of resource management skills is taking on a high level of importance as a way of ensuring that errors are effectively detected and managed (Flin & Martin, 2001; Flin et al., 2000). It should be noted that CRM training course is not important by many ship owner, it is because no strong rule or regulation issued by IMO in order to support it such as SOLAS, ISPS Code or even the STCW. We hope that through the revision of STCW according to the 2010 amendment it will adjust CRM training as mandatory course for future seafarers. In addition to the above explanation, the limitation of bridge resources or bridge equipments should be considered by seafarers. Bridge watch-keepers should be aware of the dangers of being over-reliant on these devices and:

- understand the types of errors that are possible and recognise the operational warnings that appear on the display;
- understand the limitations of the devices;
- regularly test the devices using the built-in operational test facilities.

4 MARITIME CREW RESOURCE MANAGEMET

Maritime Resource Management and Bridge Team Management Course are offered by maritime Universities in all over the world in order to concentrate on optimum approach of seafarers to reduce management error. MCRM is an essential training course for crew members of ships which approved by IMO. This training program deals with management in highly operational situations on board ship's bridges, in engine rooms etc. It defined by (Poop, 2009) as "MCRM is a course that aims to provide knowledge and a practical understanding of operative management skills". As we know majority of ship accidents or incidents are being caused by human errors, therefore the main part of the MCRM course is to review several case studies of accident and incident at sea. The following table illustrates the number of Iranian and foreign maritime accidents that rapidly raised in 2009 and reduced quickly in 2010, which was because of the standard training courses such as MCRM, BRM, BTM, etc.

At this part, there are two more or less similar ideas about the non-technical or resource management which have been stated in order to improve the required skill in crisis management.

Table 1 (Source: Iranian Port and Maritime Organization)

Maritime Accidents	2007	2008	2009	2010
Total number of accidents	93	88	116	22
Number of Iranian accidents	85	82	113	19
Number of Iranian sunken ships	36	37	52	7
number of person survived	720	656	834	226
Medical care offered to seafarers	108	119	158	47

Michael Barnett, et al (2002) stated that although there is now a general acceptance of the core concepts for the non-technical or resource management skills required for competence in crisis management, there is also an acceptance that the behaviours associated with these skills are context specific. In addition, Helmreich et al. (1998) suggest that the optimal implementation of resource management skills is dependent upon the cultural context in which they are applied. It should be noted that in order to evaluate the application of resource management skills, the assessment should be carried out within a cultural context. The International Maritime Organization's Seafarer's Training, Certification and Watch keeping Code (International Maritime Organization, 1995) are the essential resource management skills. Table *A-V/2* of this code indicates the minimum standard requires in crisis management and human behaviour skills for senior officers who are in charge of passengers in emergency situations. Majority of Maritime Universities run the Resource Management training courses which are combined as both deck and engine field. The most part of the courses are taken place by using simulator technique in order to teach the lessons by sequence casualties.

5 HUMAN FACTORS

By considering the following reports, we would be able to understand the role of human on maritime accidents: Data from research undertaken by the UK Protection and Indemnity Club (UK P&I, 1997) indicates that human error directly accounted for 58% of all shipping incidents that led to major insurance claims. The United States Coastguard (1995) stated that the human element was a root causal factor in 70% of all shipping incidents. Although not all of these incidents led to a crisis situation, all had that potential. Accepting that human error is inevitable, there is a need to understand the behaviours of effective error detection and management in order to ensure safe and efficient operations (Helmreich et al., 1998). The items related to the human factors are as follows: firstly, fitness which is related to the absence of factors cause negative effect on human performance such as regular sport and exercise or even mental factor. Secondly, use of non-prescription drugs, alcohol and extreme amounts of caffeine, etc. Thirdly, seafarers communication difficulty which is because of different languages, cultures, customs and behaviours. Fourthly, fatigue of crew and the qualification with optimum training of seafarers based on the latest STCW requirements. The definition of human factor with the relevant items has been stated by DNV (Det Norske Veritas) as follows: We believe that a central result of this analysis is the paramount importance of the human factor. In fact, in the majority of cases reviewed, the incident was due to one or more of the following: Poor crew competence, lack of communication, lack of proper maintenance, lack of application of safety or other procedures, inadequate training, poor judgment of the situation, and so forth. This general conclusion also means that many of the serious accidents reviewed might have been averted if some of the above deficiencies did not exist. It should be noted that the causes of accidents are grouped in different codes as A,B,C,D,E,F,G, the codes defined by (DNV) as "DAMA" database structure which is used for both statistical and fault tree analysis. The following comments extracted from different opinion of the five authors about the cases such as "what was the cause of an accident" or "what possible measure could have prevented it" in quantitative terms. The five more important DAMA causes were:

- F04 (8.2%): Existing routines for safety control known, but not followed.
- G02 (7.9%): Insufficient real competence (practice from occupation, waters, with equipment or suchlike).
- A01 (6.6%): Very heavy weather, natural disaster, etc.
- G07 (5.0%): Not adequate observation of own position/not plotted on charts.
- G09 (4.0%): Misjudgement of own vessel's movements (current, wind etc.).

Within each group, the most important causes were: Group A: Circumstances not related to the ship: cause A01-Very heavy weather, natural disaster etc (49.5% of the total), followed by cause A07-Operational fault with other ship (wrong manoeuvre/poor seamanship etc) (16.2% of the total) and A02-Current, wind etc led to strong drift or other manoeuvre difficulties (7.1% of the total), covering together a percentage of 72.8% of Group A cases. Group B: Construction of the ship and location of equipment on board: cause B01-The ship's structural strength not sufficient (49.1% of the total) and cause B02-The structural strength weakened by later welding jobs, corrosion etc (30.9% of the total), together coming to an 80% of Group B cases. Group C: Technical conditions concerning equipment on board: cause C09- Technical fault with equipment (34% of the total) only.

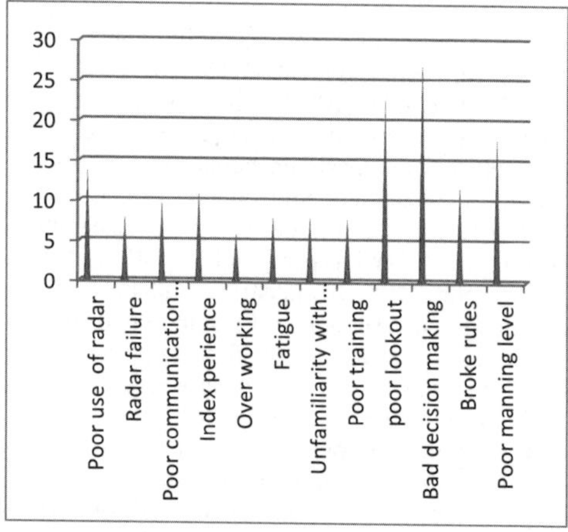

Fig.1 (Source of data: R.Ziaretie, 2010)

Group D: Conditions concerning use and design of equipment: there is no statistically significant cause. Group E: Cargo, safeguarding and treatment of cargo and bunkers: cause E01-Self-ignition in cargo/bunker, also by "sloshing" in tanks (50% of the total) only. Group F: Communication, organization, procedures and routines: cause F04-Existing routines for safety control known, but not followed (31.4% of the total), followed by cause F10-Failure of routines for inspection and maintenance on board (11.3% of the total).

Group G: Individual on board, situation, judgment, reactions: cause G02-Insufficient real competence (practice from occupation, waters, with equipment or suchlike) (22% of the total), followed by cause G07-Not adequate observation of own position/not plotted on charts (13.8% of the total), and cause G09-Misjudgment of own vessel's movements (current, wind etc) (11.2% of the total). Figure.1 shows the cause of maritime accidents and human errors as common factors at sea; it should be noted that the maximum factor is related to the bad decision making of officer on watch and the next is poor lookout of watchman, both by human errors.

By reviewing the above mentioned groups, there was a serious statistical attention which has been emphasized on groups F and G. In fact, it is not amazing; because the two groups include mostly to the maritime accidents which were due to the human error. Figure.2 shows the causes of on board ships with the main reason of bad decision and the next is poor lookout.

6 MARITIME CASE STUDY

On 6th January 2002, the Dover Strait, one of the busiest waterways in the world, was shrouded in thick fog. Visibility was less than 200 meters in places. Two ferries were crossing the Dover Strait at 0900 that day. The "Diamant" was coming from Oostende heading for Dover. The "Northern Merchant" was heading to Dunkerque from Dover. Both vessels were travelling at close to normal cruising speed: "Diamant" a high-speed craft was travelling at 29 knots, and the "Northern Merchant", a Ro-Ro ferry, was travelling at 21 knots. Were they to have continued their course and speed, the vessels' paths would have taken them to within half a mile of one another. As it was, at just over a mile apart, the bridge teams started to question the assumptions they had made about each other's probable course of action and started to implement course changes, but not speed changes, that would, they believed, put a greater distance between themselves. At 0952 they collided.

7 INVESTIGATION OF THE CASE

Seafarers Mistake: it was because of failure to maintain adequate distance; It means that 'violation of procedures' (it was too late for assuming about the actions of the other vessel in thick fog).

Seafarers Mistake: it was because of failure to reduce speed; it means that 'violation of procedures' (as a result of the speed of the craft in poor visibility, thus they had no time to take action to avoid the collision).

Seafarers Mistake: it was because of poor professional judgment; it means that 'violation of collision regulations' (As the Rule stated that, when a vessel can only 'see' another vessel on its radar and a risk of collision exists, she shall take avoiding action in ample time. Altering course at one mile or three minutes before the collision, with 21 knots speed was not in ample time).

As consequence of the report, fault of the collision was due to the Crew Resource Management or Bridge Team Management errors. Through the official report and investigation which had been carried out by the Department of Transport regarding the incident, the cause of accident was because of the operator errors (the bridge team or the Master errors) by the eighteen reasons as the official report (MAIB, 2003; pp. 43-44).

8 THE STCW CONVENTION

The Manila amendments are the result of nearly five years of intensive debates and discussions on various accessions also at the annual IMLA Conferences and at IMEC gatherings. Although the outcome is not the optimum, it is, however, an acceptable and practicable instrument suited to further develop Maritime English as an essential but relatively new knowledge area in order to satisfy the new provisions and thus the complex requirements of maritime industry. As Prof.Trenkner (2010) emphasized on the outcome of the Manila Conference which was not an optimum result, further Prof.Ziaretie (2010) had more or less similar idea in this field. It should be noted that several attempts carried out in order to revise STCW95 convention through the Manila Conference 2010. Nevertheless, the implementation of the revised Convention will take time and a number of deficiencies kept in the Convention until the proposed changes to be taken place.

In addition to the above explanation, regarding the role of optimum training of MET based on STCW Convention Prof.R.Prasad (2010) stated that seafarers need to acquire comprehensive understanding of technical facts through active learning processes that enhances deep understanding of scientific as well as social concepts and help develop technical, cognitive and social skills. Skills for group/team work, good communications and resolving issues are as essential because they have to work in such environment.

9 THE MARCON PROJECT

The MARCON is an improvement of maritime lecturers' competencies project which is based on Lisbon European strategies fortune in order to maintain the European maritime university system as worldwide framework. It should be noted that the eEurope has been launched as e-learning in 2001, therefore communication and computerized technology became the main component of maritime training. The objective of this project has been stated by Prof.R.Hanzu-pazara et al (2009) as follows: The general objective of the MARCON project is represented by multidisciplinary research concerning initial and continuous formative of the lecturers from maritime universities and providing of advancement programs according with the maritime industry requirements.

Although, around 40% of the present teaching staff of Constanta Maritime University has pursued this program, therefore the author believes that if all maritime lecturers become familiar with the latest technologies, advance computerized system and simulation procedure, than their competencies in maritime training will be improved in a great extent.

10 CONCLUSIONS

Many maritime accidents have been reported in the last couple of decades, there were because of the shortage of seafarers' skill to supervise both resources and crises. CRM training has been seen gradually more as a fundamental part of the human error management viewpoint. The International Maritime Organization gave the impression of the require for resource management skills on board ships, nevertheless the standards of competence and their evaluation criteria are not fully formed similar to the civil aviation. CRM training is a technique that has been formulated for organizing people to arrange maritime/bridge resources in order to avoid collision at sea.

REFERENCES

Barnett, M. L., Gatfield, D., and Habberley, J. 2002. Shipboard crisis management: A Case Study. Proc Int. Conf. Human Factors in Ship design and Operation. pp 131-145 RINA, October 2002

Cardozo, D. 1993. Crew Resource Management, Journal of Management Development 12 (6): 7-14.

Cherrington.D 1995. The Management of Human Resource, Englewood Cliffs, NJ: Prentice-Hall.

Cole.C, Trenkner.P. 2010. International Maritime lecturer Association, proceedings of IMLA18 Conference, Shanghai – China.

Cross, S.J. 2010. International Maritime lecturer Association, proceedings of IMLA18 Conference, Shanghai – China.

Edkins, G. D. 2002. A review of the benefits of aviation human factors training.

Flin, R., & Martin, L. 1998. Behavioural markers for crew resource management. UK Civil Aviation Authority Paper 98005. London: Civil Aviation Authority.

Flin, R., & Martin, L. 2001. Behavioural Markers for Crew Resource Management: A Review of Current Practice. The International Journal of Aviation Psychology, 11(1), 95-118.

Gyles and Salmon 1978. Recent Developments in Crew Resource Management (CRM) and Crisis Management Training, Southampton, UK.

Gyles J.L. & Salmon, D. 1978. Experience of Bridge team Training using the Warsash Ship Simulator. Proc First In Conf. on Marine Simulation MARSIM 1978. pp 1-26 Nautical Institute.

Hanzu-Pazara.R. et al. 2009 as follows: The general objective of the MARCON project.

Helmreich, R., Wilhelm, J., Klinect, J., & Merritt, A. 1998. Culture, Error and Crew Resource Management. Austin: University of Texas at Austin.

International Maritime Organization. 1995. Seafarer's Training, Certification and Watch keeping Code (STCW Code). London: IMO.

Marine Accident Investigation Branch. 2003. Marine Accident Report 10/03. Report on the Investigation of the collision between Diamant/Northern Merchant. Southampton: MAIB.

Miler.W. 1987. Human Resource Management. Upper Saddle River, NJ: Prentice-Hall.

Prasad Rajendra, 2010, International Maritime lecturer Association, proceedings of IMLA18 Conference, Shanghai – China.

Royal Navy. 2002. Basic Operational Sea Training Assessment Form. Plymouth: HMS Drake. Flag Officer Sea Training.

United States Coastguard 1995. Prevention through People, Quality Action Team Report. Washington D.C.

UK P&I Club .1997. Analysis of major claims – ten-year trends in maritime risk. London: Thomas Miller P&I Ltd.

Ziaretie.R. 2010. International Maritime Lecturer Association, proceedings of IMLA18 Conference, Shanghai – China

2. Women Seafarers: Solution to Shortage of Competent Officers?

M. Magramo & G. Eler
John B. Lacson Foundation Maritime University, Iloilo City, Philippines

ABSTRACT: This study aimed to determine the participation of women in the seafaring profession. It also looked into the hiring practices of the different manning and shipping companies in the country. It also tackled the hindrances or obstacles a woman seafarer faces in a male dominated world like seafaring. This research utilized the interview approach in data-gathering among the crew managers, and an in-depth interview with a lone woman seafarer participant.

1 INTRODUCTION

Throughout history, women have always aimed for a recognized place in society. Society has prescribed that a woman's place is usually at home, taking care of the children. Many feminist movement forced the issue of women's rights to come into people's awareness These frontrunners have helped redefine and consolidate the nature of women's contributions to society shows that progress has been made.

In today's society women have access to education and can promote themselves much more easily than in the seventies. Women's changing role is happening because women nowadays are educated.

For centuries, maritime history and literature have treated seafaring as a solely male domain. Although women have now begun to appear in maritime scholarships, they are mostly on the periphery. A few women have been recorded as having travelled as stewardesses , explorers, or companions to captains, but on the whole, women did not take part in the actual running of ships(Creighton and Norling, 1996; Stanley, 1987 in P. Belcher, 2003)

Women on boardships either served as children's nurses, stewardesses for women passengers and as laundresses.

It was only after 1945 when women were regularly recruited as stewardesses, cooks and radio officers. The first women cadets were recruited only in the late 60s.

Since the late 1990s there has been a growing interest in training and recruiting women seafarers. This is largely connected to perceived shortages of officers in the world fleet.

The latest Baltic and International Maritime Council (BIMCO) and the International Shipping Federation (ISF) report suggests that the current shortfall of officers corresponds to 4 percent of the total (16,000 officers) and predicts a 12 percent shortfall by 2010, an estimated 46,000 officers.

As of 2006, the European Commission reported an estimated 36,000 officers shortfall from the 13,000 officers in 2001(P. Belcher, et. al, 2004).

Owing to the fact that there is big shortage for officers to man the world's fleet, the focus now is on the possibility of recruiting more women into the maritime profession.

The United Nations promoting women's employment and the integration of women into all levels of political, economic, and social development the IMO produced a strategy for integrating women into the maritime sector in 1988 , when it began to implement its Women in Development (WID) programme concentrating on equal access to maritime training through both mainstream programmes and gender specific projects. One of the immediate impacts of the programme has been the rise in the percentage of women students taking part in the highest level of maritime training.

The SIRC/ILO survey in 1995 revealed that women made up of less than 8 percent in the total umber of students at the World Maritime University. Overall, the participation rate of women in seafaring remains low. Only about 1 to 2 percent of the world's 1.25 million seafarers and that most of these women are form developed countries. The study by Belcher, et al.(2003) found out that women continue to constitute a very small part of labor force of seafarers. It is in this context that this study was conducted.

2 STATEMENT OF THE PROBLEM

This study aims to determine the participation of women in the seafaring profession in the Philippines. Specifically, it aims to determine the following:

1. The number of Filipino women currently employed on board vessels;
2. Hiring practices and policies of the shipping companies in recruiting women for seafaring;
3. Reasons why some ship owners refuse to hire women on board ships

3 METHOD

This study employed the qualitative method of research utilizing the interview approach in gathering data. It aimed to determine the participation of women in the maritime profession. Likewise, it looked into hiring practices of the ship owners and it also intended to determine the reasons why many ship owners refuse to take or hire women in their company. The participants in this study are the shipping and manning companies and a woman seafarer who had been at sea and is now an officer.

4 RESULTS

4.1 *Modes of analysis*

On December 9-13, 2007 these group of researchers proceeded to Manila to conduct interviews with the different stakeholders in the maritime profession. The interview focused on different issues confronting the maritime profession. In the light of the many issues on the shortage of officers, the women remain untapped as a possible source of the qualified seafarers. The first interviewee was Mrs. Virginia Linesis, the lady president of K-Line. She was quick in saying that as far as her company is concerned, they are not hiring women as seafarers in their company. She also told the group that Japanese culture does not allow women to work on board their vessels.

Capt. Jose De la Peña also shares the opinion of Mrs. Linesis and other ship manning managers with Japanese principals do not hire women seafarers.

C/M Erickson Pedrosa said his company, the Walllem Maritime Services are employing women in the company. He further said: "We have five (5) women cadets. This company is giving fair treatment to men and women. The entry of women in the profession will help in the shortage. However, women are weak emotionally and physically. They must be trained just like the men and be prepared for the job.

Capt. Jimmy Milano, the General Operations Manager of Inter-Orient Maritime Enterprises, Inc. said they don't employ women on cargo vessels, but they have at least two (2) women stewards in the company's pleasure yacht numbering about 14. Personally, though, he is against women employment on cargo vessels with mixed crew because it could sometimes spell trouble. But he is also in favor of hiring women on passenger vessels.

Capt. Rolando Ramos of Unisea, Philippines believed that women are capable of becoming officers and doing the work on board vessels, however because of our culture, women are easily tempted. The company does not discriminate women as a matter of fact, the company employs women.

3/M Glena Juarez, a graduate of JBLFMU is now a licensed Third Officer. She sits as the Assistant Technical Marine Superintendent at UNISEA, Philippines, Inc. She believes that women are good in planning procedures and office work. Women are different in handling work on board. She adds "all the things we do are arranged. We have a system of doing things."

Asked if she would still continue life at sea despite the fact that she now holds a very comfortable and a very important position in the office, she smiled and said "Yes, for me it's not fulfilling that I have the 3rd Mate license but I did not have the chance to practice." The owner's representative made her choose between a land based and a sea-based position. The owners plan to put a junior third officer on board is a bit longer and for a while she opted to stay in the office because she also needs to know how the office runs, how the shipping is managed ashore. Asked how long does she intend to stay on board. She said: "I don't want to stay on board all my life. It's difficult to stay on board.

When asked what makes life on board difficult for someone like her, she said: "I guess the environment, the weather. The bad weather, you cannot sleep because you are bothered by the rolling and pitching of the vessel. And also working with people on board, there are people who are mean! The entire working environment makes life on board difficult. She experienced 36 hours of no sleep because they have to finish the cargo hold cleaning, because after 6 o'clock we cannot discharge the cargoes. She also said: "There are times I wanted to go home because of the people on board." Even the people from the engine department would ask her why she joined the profession."

When asked how she was able to finish the one year contract on board, she mused: I have to prove that I can survive. The workload is not a problem, I guess because it's part of being a seafarer. To her, one must be aware that sleep on board is not the kind of sleep you are experiencing on land. To her, while you are sleeping your mind must be awake because what if there is an emergency. Alertness is always there. You cannot say that your sleep is a rest, you are still awake. For her, just one ring of the phone she is already awake. She needs to be always alert. There are also times its fun to be on board having the chance to visit other places. She also said that she is proud whenever the captain asks her to clean the bilge, the strum box or rose box.

Asked if she was able to apply what she learned in school, she was quick to say: "Yes. Especially navigation, it was more enhanced on board especially the skills. I guess you have to love your work in order to survive."

"During my free time, I just stayed in my room watching movies or sometimes talking with other crew members. Every time there are nasty jokes, I just listen or just ignored it. Sometimes, I simply get out of the room or cabin to avoid embarrassment or being offended by the conversation.

Asked what her job now is in the office; she said: "I am the Assistant Technical Superintendent. Together with the Technical Superintendent, we compose the Ship Management Team. It's a new team in the company; we have a counterpart in the main office in Greece. We visited the main office of UNISEA in Greece last October 2007. We observed how the office is run, how the office operates.

Here in the Philippines, 3/M Glena also works in the administration department of manning. The Greek Director told her she is being prepared either as Human Resource Officer or Quality Management Representative in the office after her training on board and in the office.

Capt. Martinez believes that Filipino women are capable of being officers on board vessels. He believes that if there are Danish or British women officers, so can there be Filipinos. He believes that these women must learn to love this job, because it's a tough job.

Mrs. Brenda Panganiban, President of Bouvet Shipping Management, said she is in favor of women seafarers in the profession but only for a very limited time. This is because she will be getting married. European principals accept women in the profession but not in Japanese owned vessels. She also mentioned that women have different disposition when it comes to decision making. This opinion is also shared by the training manager of Maritime Corporation, Capt.Lexington Calumpang. According to him "women are for light work only. For the deck there is no problem because they are on the bridge, but in the engine their strength must be like that of the men. For theory women are good. In this company, hiring women is preparatory to office job after their training on board is completed. We have a woman superintendent from MAAP. She is now the safety environmental superintendent with a 4th engineer license. We have employed two women who just disembark; they were taken by the company two weeks after they have disembarked." He further stressed that on deck women are slow in decision making. They rattle. One woman was given penalty for a year; she is assigned at the office of the principal's representative in the training department for the meantime.

When asked if they have any policy requirement for women, they too, like men have to pass the admission exam, interview and the medical examinations. They also have to undergo training both in-house and in other training centers.

Women on board vessels are usually visited by the fleet manager to see to it that they are not victims of any form of harassment from the other crew. Women seafarers must perform just like the men, when they become a master mariner they can be assigned in the office.

When asked what her opinions are on women seafarers, Mrs.Carla Limcaoco, Executive Director and Vice- Chairman of Philippine Transmarine Carrier said: "First of all, these are issues of physical demands. Second, women will eventually fall in love, get married and get pregnant. But if they are focused and determined enough then there is no problem."

5 FINDINGS

1. Women seafarers in the seafaring profession are comprised of only 1-2 percent.
   Most of these women are in the steward department although there are a few ship owners who are now hiring women as officers.
2. Not too many of the companies in the country today are hiring women to become officers. The companies hiring these women are actually preparing them for office positions. Women are more organized; hence they are more appropriate to work in the office after they have acquired the necessary training and knowledge of running a ship.
3. The women who are now taken as cadets and being prepared to become officers on board ships must be physically, emotionally and spiritually tough in order to overcome all kinds of hindrances, obstacles and challenges that may come their way. They need to think and act properly in a profession that is male-dominated.

REFERENCES

Belcher P. Sampson H., Thomas M. Viega J. Zhao M., 2004. Women Seafarers: *Global employment policies*, ILO, ISBN 92-2-113491-1, UK

BIMCO/ISF, 2005, Manpower Update: *The Worldwide Demand of and Supply for Seafarers Institute fro Employment Research, University of Warwick.*

Seafarer's International Research Centre (SIRC 1999). *Seafarers*. Cardiff University, Wales UK

Skei, O. M. (*The growing shortage on qualified officers),* a paper delivered at the 8th Asia-pacific Manning and Training conference, November 14-15, 2007. Manila Philippines.

3. The Manning Companies in the Philippines Amidst the Global Financial Crisis

M. Magramo, L. Gellada & T. Paragon
John B. Lacson Foundation Maritime University, Iloilo City, Philippines

ABSTRACT: This qualitative research looked into the effects of the global financial crisis to the shipping and manning industry in the Philippines. It utilized the focus group discussions and the in-depth interview of the company representatives of the shipping companies and crewing managers of the different manning companies in the Philippines. Findings of the study revealed that the bulk carriers, container vessels, car carriers and cruise vessels are the type of vessels directly affected while tankers (crude, gas, chemical) are slightly affected. It was found out that the shipping companies with long-term contracts with their charter parties are not affected by the global financial crisis. On the other hand, those players with no long-term contracts are greatly affected. The situation also somehow eased the global shortage of competent officers in the manning industry.

1 INTRODUCTION

The global financial crisis started to show its effects in the middle of 2008. Around the world stock markets have fallen, large financial institutions have collapsed or been bought out, and governments in even the wealthiest nations have had to come up with rescue packages to bail out their financial systems (http://www.globalissues.org/article/768/global-financial crisis).

In this same article, it was mentioned that the collapse of the US sub-prime mortgage market and the reversal of housing boom in other industrialized economies have had a ripple effect around the world. Other weaknesses in the global financial system have surfaced. Some financial products and instruments have become so complex and twisted, that as things start to unravel, trust in the whole system started to fail.

The extent of this problem has been so severe that some of the world's largest financial institutions have collapsed. Other have been bought out by their competition at low prices and in other cases, the governments of the wealthiest nations in the world have resorted to extensive bail-out and rescue packages for the remaining large banks and financial institutions.

In Europe, a number of major financial institutions have failed, or needed rescuing. A number of European countries are attempting different measures as they seemed to have failed to come up with a united response. Some nations have stepped in to nationalize or in some way attempt to provide assurance for people. This may include guaranteeing 100% of people's savings or helping broker deals between large banks to ensure there is no failure.

For the developing world, the rise in food prices as well as the knock-on effects from the financial instability and uncertainty in industrialized nations; are having a compounding effect. High fuel costs, soaring commodity prices together with fears of global recession are worrying many developing country analysts according to Kanaga Raja (UNCTAD, Third World Network, September 4, 2008).

Countries in Asia are increasingly worried about what is happening in the West. A number of nations urged the United States to provide meaningful assurances and bailout packages for the US economy, as that would have a knock-on effect of reassuring foreign investors and helping ease concerns in other parts of the world.

As the turbulence of recent weeks has rocked the previous complacency that had become embedded in the fabric of a western society that the righteous have already dubbed as irresponsible, the marine industry has begun to batten down in preparation for rough weather ahead.

With the consumer world fattened itself on the produce of its spending, shipping, too, enjoyed a smooth voyage while it supplied the needs of the spenders, automatically reaping the benefits bestowed by the boom.

Storms are part of the seafarers' routine and the cyclical nature of the industry has been present since the early days of steam- a time when the fleet ship owner came into his own. Over the years, times have been good and times have been bad and so the industry learned to adapt itself, accordingly by expedient expansion or contraction, as a direct function of consumer demand.

Stormy waters are always present somewhere and by skillful navigation they can often be avoided. Some would argue there is no difference – this predicted world- recession is like previous downturns- but therein is the answer. It is likely to be a world-recession – not just the United States, not just Europe – the globalization factor is the difference.

With indicators in China now predicting a slowdown, the source of much of shipping's recent success may be faltering, thereby exacerbating the impact to be felt by many operators. The extent of such an impact is still a variable since it shall be dictated by the actions of the banking sector- particularly those concerned with ship finance of course.

Shipping has always been capital intensive and as such is very reliant on the banks to provide the buoyancy needed to operate smoothly and successfully. The extent to which the banks now restrict their lending will therefore be directly proportional to the difficulties that shipping companies may or may not be about to experience.

At the present time, "caution" is the dogma that the bankers are muttering in their hypocrisy so it is certain that prudence will prevail. According to some observers, lending is not likely to dry up, but they are also saying that conditions have already been made considerably more stringent with risks being minimized and borrowing limits curtailed (Kennedy, F. (2008). *Shipping will ride the storm of financial crisis,* http://www.gulfnews.com/business/shipping).

It has also become evident that bank guarantees are also becoming difficult to obtain which will have a particular impact on shipbuilding- however, in the long term perhaps that maybe beneficial in view of the predicted glut in tonnage that will certainly not help a recovering industry after this present crisis. At the end of the day the strong will survive and the weak will fail and the fact that much of the severity of the situation for the ship operator rests with a bunch of bankers, is an understatement.

Notwithstanding the gloom, it must be remembered that the shipping industry is resilient – a fact mould into its very fabric by its cyclical nature mentioned earlier.

Furthermore, the momentum of the shipping industry will never stop due to its place as the cheapest form of transportation. It will certainly slow down as we now experience, but recovery will be the next rend and likely to be 'in-line' with the economic performances of the main-drivers of the 21st century- namely China and India.

But for the time being, contraction is the scenario-while times have been good, the interdependence of all economies on each other has suited shipping.

As a global industry it has provided a global service to a global economic structure in order to satisfy all requirements- worldwide and local.

It is on this premise that the present study was conducted.

2 OBJECTIVES OF THE STUDY

This study was conducted to ascertain the effects of the global financial crisis to the shipping and manning companies in the Philippines. It further looked into what measures are being undertaken by the individual shipping and manning companies in the country today and what type of vessels are directly and indirectly affected.

3 METHOD

This descriptive qualitative study utilized the phenomenological approach in data gathering. It also made use of data analysis of literature published in refutable journals.

Basically, it utilized the interview and the focused group discussion techniques in data gathering.

3.1 *Participants of the study*

The participants in the study were the twelve (12) managers of the shipping and manning companies in the Philippines. Most of the companies were based in Manila and the interview was conducted from December 6-13, 2008.

3.2 *Data Analysis*

3.2.1 *Effect to dry cargoes*

According to Capt. Dag Ulfstein of KGJS Manila, their company is not directly affected by the global financial crisis because "the company has a solid financial base." The company (KGJS) has existing long term contracts with charterers and we are confident we will be able to surpass this. When the freight were skyrocketing, the company did not benefit on these because of our long term contracts. Other companies will definitely suffer a bit. We heard now that in Subic, there are up to 15 ships being laid off, there are ships laid off in Davao, and we have just heard today that in Busan, Korea ships from the yard ready for delivery to the owners are just being towed outside the port without the crew. And these are brand new ships, they are going straight from the yard to lay off. They are not taken into service. Maersk, for example laid out six ships already, probably at least 15 ships later. As I said earlier in Subic MOl also has six bulk carriers, probably 15,000 tonners laid out there and they heard all over the world and Hongkong is first, there is no more space in Singapore. There are ships being laid off everywhere. So for some companies who don't have solid financial base and no long term contracts, they will definitely suffer. Expect some companies will go down (and probably will declare bankruptcy) the big ones will still be there. They will just slow down a bit on their recruitment and look at their spending but serious companies will have to remain in the shipping industry. They will continue, maybe a bit easier for the coming years but they will never stop on training. The impact of the global financial crisis to a company like KGJS has been minimal. They have not been gambling much lately on the market because they have a solid base contract or a long term contract. Maybe 20 percent of the bulk carrier will be laid off and old vessels will be scrapped. Car carrier will also be affected as many people would probably not buy new cars because of the economic slow down. This was also confirmed by Capt. Remigio Zamora of Odfjell. He also said that the dry cargo carriers, bulk cargo, containers, the car carriers and to a certain extent passenger ships are the type of ships that were directly affected by the global financial crisis. Cape size bulk carriers are the ones badly hit by the global financial crisis. Freight rates has gone down as low as 50 percent of the original cost according to Capt. Ramos. At UNISEA, they have a vessel with almost 70% decrease on the cost of freight rate just for the vessel to be able to return to Asia.

3.2.2 *Effect to tankers*

Certain tanker sizes will be affected but not to the same level as the dry cargo carriers. According to Capt. Remegio Zamora of Odfjell, the tankers are not that affected by the economic downturn. JO tankers, Stolt and Odfjell are manning companies and at the same time ship owners, so we are not affected. The managers of the mentioned manning companies were one in saying that oil tankers will continue to do its usual business. Oil and other oil-based products including chemical and gas carriers will continue with its normal or usual business through out the world.

3.2.3 *Effect to dry cargoes*

According to Capt. Dag Ulfstein of KGJS Manila, their company is not directly affected by the global financial crisis because "the company has a solid financial base." The company (KGJS) has existing long term contracts with charterers and we are confident we will be able to surpass this. When the freight were skyrocketing, the company did not benefit on these because of our long term contracts. Other companies will definitely suffer a bit. We heard now that in Subic, there are up to 15 ships being laid off, there are ships laid off in Davao, and we have just heard today that in Busan, Korea ships from the yard ready for delivery to the owners are just being towed outside the port without the crew. And these are brand new ships, they are going straight from the yard to lay off. They are not taken into service. Maersk, for example laid out six ships already, probably at least 15 ships later. As I said earlier in Subic MOl also has six bulk carriers, probably 15,000 tonners laid out there and they heard all over the world and Hongkong is first, there is no more space in Singapore. There are ships being laid off everywhere. So for some companies who don't have solid financial base and no long term contracts, they will definitely suffer. Expect some companies will go down (and probably will declare bankruptcy) the big ones will still be there. They will just slow down a bit on their recruitment and look at their spending but serious companies will have to remain in the shipping industry. They will continue, maybe a bit easier for the coming years but they will never stop on training. The impact of the global financial crisis to a company like KGJS has been minimal. They have not been gambling much lately on the market because they have a solid base contract or a long term contract. Maybe 20 percent of the bulk carrier will be laid off and old vessels will be scrapped. Car carrier will also be affected as many people would probably not buy new cars because of the economic slow down. This was also confirmed by Capt. Remigio Zamora of Odfjell. He also said that the dry cargo carriers, bulk cargo, containers, the car carriers and to a certain extent passenger ships are the type of ships that were directly affected by the global financial crisis. Cape size bulk carriers are the ones badly hit by the global financial crisis. Freight rates has gone down as low as 50 percent of the original cost according to Capt. Ramos. At UNISEA, they have a vessel with almost 70% decrease on the cost of freight rate just for the vessel to be able to return to Asia.

3.2.4 *Effect to tankers*

Certain tanker sizes will be affected but not to the same level as the dry cargo carriers. According to Capt. Remegio Zamora of Odfjell, the tankers are not that affected by the economic downturn. JO tankers, Stolt and Odfjell are manning companies and at the same time ship owners, so we are not affected. The managers of the mentioned manning companies were one in saying that oil tankers will continue to do its usual business. Oil and other oil-based products including chemical and gas carriers will continue with its normal or usual business through out the world.

3.2.5 *Effect to manning*

The global financial crisis has also a minimal impact to the shortage of competent officers to man vessels. As of December 2008, it's too early to really assess the impacts of the economic downturn to the shipping industry especially to manning. Some sectors say it is nature's way of balancing things. The sky rocketing wage wars among companies will now stop because some ships will no longer operate. But this will entail a lot of training especially on the senior officers who have specialized on certain types of ships. A senior officer from a bulk carrier will find it difficult if not impossible to assume senior positions on tankers. According to Ms. Minerva Alfonso, the shortage of competent officers and the lesser demand for officers brought about by the financial crisis will balance out, they will even out. The competitive edge will be competitive enough to be able to compete with other crew elsewhere. So, I think it's basically some companies will retain their preference of seafarers, it's not just the money, you also have corporate cultures, you have loyalty, you have several other factors. Captain Remegio Zamora also believed that the global financial crisis will level off the problem of shortage of competent officers. A year ago, when the demand for officers was so great, the name of the game is highest bidder. For the officers who have this quality of service or experience, they will demand for higher salaries. This is bad for the industry because loyalty is gone. It is all about money. But with the crisis, the name of the game now will be loyalty check. Yes, this will level off maybe in the next four years.

3.2.6 *Effect to Filipino Seafarers*

In the light of the global financial crisis, the Filipino seafarers will still be the preference of ship owners. Asked if the foreign ship owners will start looking for cheap labor in other countries, Capt. Zamora said being in the business for so long now, having attended international seminars and conferences and having talked to so many principals, he believed that the Filipino seafarers will continue to be in demand. In an economic crisis like this, the focus of any company is to save money. In shipping, one qualified and competent seafarer equals five new seafarers. It had been proven in the past, a Filipino officer is better than four Chinese officers. A Filipino officer may cost US $2,000 while a Chinese counterpart will ask only US$ 500.00. But the quality of work of a Filipino is much better than four Chinese. "We Filipinos are flexible and we can communicate, we are courteous, we are polite, and very hardworking." These are Filipino traits that when compared to other nationality is considered as our edge or advantage.

4 FINDINGS

1 The global financial crisis has directly affected dry cargoes especially the cape size bulk carriers. For companies with no long term contract, they are left with no other option but to lay off their vessel.

2 The tanker vessels on the other hand are not really feeling the crunch of the economic downturn as the demand for oil and chemical and gas worldwide will continue to be present.

3 The shortage of competent officers has somehow ease out due to the crisis; many newly constructed ships are just being towed from the yard to the port. Otherwise the need for officers to handle new ships would have been almost impossible to find.

4 Some ship managers in the Philippines see this as an opportunity to manage more ships that the ship owners themselves are unable to do.

5 The Filipino seafarers will continue to be the officers and crew of preference among the ship owners. With the economic downturn, ship owners need to cut down on crew cost, and the preference of these ship owners will always be Filipinos despite the relatively higher salary rates among other Asian officers and crew.

REFERENCES

Kanaga Raja (UNCTAD, Third World Network, September 4, 2008).

Kennedy, F. (2008). *Shipping will ride the storm of financial crisis,*

http://www.gulfnews.com/business/shipping).

http://www.globalissues.org/article/768/global-financial crisis).

4. Academe and Industry Collaboration: Key to More Competent Officers?

M. Magramo, G. Eler & L. Gellada
John B. Lacson Foundation Maritime University, Iloilo City, Philippines

ABSTRACT: This descriptive-quantitative-qualitative research looked into the level of work performance on board among the graduates of a maritime university. It also looked into what skills and qualities the graduates need to enhance in order to conform to the demands of the shipping and manning industry. It further sought to identify the strengths and weaknesses of the academic curricular and co-curricular programs of the university in order to meet the standards of the industry. The respondents of the study were the company representatives, crewing and personnel managers from the 19 shipping and manning companies in the Philippines. It utilized a validated survey instrument in gathering quantitative data. The interview and focused group discussion (FGD) were utilized in gathering the qualitative data. The quantitative data showed that the over-all performance of the graduates was rated very good workers by the participants of the study. The graduates have no attitude problems and they are well adjusted with their life on board and that they can easily get along with co-workers. This was affirmed by the very good rating on teamwork. The graduates have good potential for leadership, strong work ethics, and can be relied upon whenever they are assigned to do a particular task. Despite the positive results obtained in the quantitative data collected, the qualitative data showed otherwise.

1 INTRODUCTION

The maritime schools today are faced with the challenge of producing graduates who are competent enough to man foreign going vessels. But the quality of these graduates coming from the maritime schools is always questioned by the shipping companies. Many of these graduates are not academically prepared or do not possess the skills needed for the job on board ships.

According to the shipping industry the current curriculum does not meet the needs of the industry.

It is noteworthy that in some instances maritime schools are not capable of acquiring state of the art equipment that will better facilitate transfer of knowledge and skills badly needed by the technology-driven shipping industry.

Academe and industry linkage in the Philippines at this stage is still in its infancy, not too many industry are extending assistance or in collaboration with the different maritime institutions today.

According to del Rosario (2007) there is a need to enhance industry-academe linkage to improve the quality of lives of our graduates, That is, "by making the Philippine education system competitive, effective and efficient, universal and inclusive."

Further, del Rosario stressed that "the time has come to think about recreating the way we learn and do away with traditional divides".

Atlan (1990) and Peters and Fusfeld (1982) in Wu (1994), several reasons motivate the industry to increase university-industry cooperation. They are: (1)access to manpower; (2) access to basic and applied research results from which new products and processes will evolve; (3) solutions to specific problems or professional expertise, not usually found in an individual firm; access to university facilities, not available in the company; (5) assistance in continuing education and training; (6) obtaining a prestige or enhancing the company's image; and (7) being good local citizens or fostering good community relations.

On the other hand, the reasons for universities to seek cooperation with industry appear to be relatively simple. They are: (1) Industry provides a new source of money for university; (2) industrial money involves less "red tape" than government money; (3) industrially sponsored research provides student with exposure to real world research problems; (4) Industrially sponsored research provides university researchers a chance to work on an intellectually challenging research programs are available; (5) Some government funds are available for applied research, based upon a joint effort between university and industry.

In the study of WU(1994), it was found out that even in Taiwan, there is still a need to strengthen the linkage between the university and industry to enhance the level of technological sophistication.

A study conducted by Santoro (2000) stresses that industry-university alliances can be instrumental in facilitating the industrial firm's advancement of both knowledge and new technologies.

JBLFMU, as a university maintains a strong partnering relationship with the shipping industry who acts as the voice of the customer, articulating the mandated competencies and necessary innovations on the academic operations of the school. The representatives from the shipping and manning agencies are invited to participate in local, national and international maritime conferences and forums hosted by the university to identify trends, needs and concerns of the maritime industry. JBLFMU has established tie-ups with 31 shipping companies for student scholarships and 74 shipping companies and agencies for the placement of cadets.

It is for these reasons that this study was conducted.

2 STATEMENT OF THE PROBLEM

This survey was conducted to attain the following objectives:

1 To determine the level of work performance on board among the graduates of the university.
2 To look into what additional skills and qualities the graduates need to enhance in order to conform to the demands of the shipping and manning industry.
3 To identify the strengths and weaknesses of the academic curricular and co-curricular programs of the university in order to meet the standards of the industry.
4 It also attempted to determine present collaboration initiatives between academe and industry.

3 METHOD

The participants in this survey were the company representatives, crewing and personnel managers from the 19 shipping and manning companies. The questionnaire was administered personally by the researchers.

3.1 *Dimensions of evaluation:*

The stakeholders evaluated the performance of JBLFMU graduates in the following categories: 1) educational dimensions (16 aspects), 2) education and training needs for further enhancement, and 3) comments and suggestions of shipping executives

This survey was conducted last December 7-10, 2009 in Manila with 19 shipping company presidents, representatives, managers, training directors and personnel managers. It utilized both the quantitative and qualitative method of research. The quantitative data were gathered using the instrument prepared by the office of the Academic Director. Individual interviews and small group discussion were likewise employed in gathering the qualitative data for this survey.

4 RESULTS

Graduates Level of Work Performance

Educational Dimension	Mean	Verbal Interpretation
Ability to work with others	4.04	Very good
Potential for leadership	3.92	Very good
Teamwork	3.84	Very good
Reliability	3.76	Very good
Ability to take initiative	3.76	Very good
Strong work ethics	3.76	Very good
Competencies prescribed by STCW'95	3.72	Very good
Job-relatedconceptual knowledge	3.71	Very good
Computer skills	3.68	Very good
Ability to correct criticism	3.56	Very good
Willingness to take risk and responsibility	3.56	Very good
Time management	3.48	Very good
Communication skills	3.44	Very good
Technological competence	3.44	Very good
Job-related technical knowledge	3.4	Very good
Critical thinking/problem solving abilities	3.32	Average
Over-all work performance	3.62	Very good

Mean scale	Descriptive level
4.21-5.0	Excellent
3.41-4.20	Very good
2.61-3.40	Average
1.81-2.60	Good
1.00-1.80	Poor

4.1 *Quantitative Data Analysis*

When taken as a whole the evaluation of the performance of the graduates of JBLFMU is very good with a mean of 3.62.

The graduates were rated 4.04 in their ability to work with others. This implies that the graduates of JBLFMU have no attitude problems and they are well adjusted with their life on board and that they can easily blend with co-workers. This was affirmed by the very good rating on teamwork (3.84).

The graduates also possessed very good potential for leadership with a mean of 3.92. This implies that the shipping managers have seen that the graduates whenever given the chance to lead will always come out as leaders.

Strong work ethics are very evident in the graduates of the university (3.76). This can be interpreted as strong work values and this is further supported with a very good rating on reliability (3.76). When graduates of the university are given a task to be accomplished, they can be relied upon to carry out the task. They were also rated very good (3.76) in their ability to take initiative. Again, this implies that the students are also resourceful and do not rely on the usual procedures or process of doing things.

The graduates of JBLFMU are competent as far as the competency prescribed by STCW '95 is concerned with a rating of 3.72. This was supported by the very good rating on job-related conceptual knowledge with a mean of 3.71.

Job-related technical knowledge, technological competence, computer skills are some of the qualities the graduates already possessed.

However, the critical thinking skills and problem solving abilities of the students were rated average (3.22). It is presumed therefore that maybe in some circumstances some of the students did not take time in analyzing deeply the problem at hand.

4.2 *Qualitative Data Analysis*

Need to enhance skills in mathematics- It was found out that some graduates of the university find it difficult to solve simple problems in finding distance speed and time and even on the conversion of units of measurement. There is a need for teachers to emphasize the use of long hand mathematics rather than relying on the use of calculators especially on sine, cosine law and Napier's rule and its application to Navigation and Seamanship to ensure the enhancement of skills in mathematics.

Need to address some students with personality problems- This was also raised by some training managers that there are some instances where they encounter students who are stubborn or hard-headed. Sometimes interpersonal problems arise brought about by many factors the likes of culture, upbringing, value system, to mention a few.

Need to enhance the communication skills of some students. During interviews some graduates are requesting if they can speak in " Ilonggo" or "Bisaya." It has been that there is very narrow range between the average and very good rating in this aspect quantitatively. Many of the participants in this study emphasized the need to continually improve the communication skills of the graduates especially on reports both in written and oral communication.

Need to emphasize the value of loyalty to the company. Some training managers complained that there are graduates of the university who transferred to another company in favor of higher salaries. Moreover, there are also officers who transferred to other companies when they have already practiced their license as master mariner in their company. The issue of loyalty is difficult to assess when financial consideration is involved.

Need to emphasize their career path especially among the scholars. Graduates of the university must focus on their career in the future. They should not be contented with the present salary with their comfort zones. Some graduates are just happy being ratings or in some cases simply as boatswain.

Emphasize discipline as a virtue. In some instances, there are students who seem arrogant, although others are too timid or too shy in talking to authorities.

4.3 *Strengths of the graduates of the university:*

1 Quantitative data showed very good performance in the different aspects on educational dimension.
2 The graduates of JBLFMU have no attitude problems and they are well adjusted with their life on board and that they can easily get along with co-workers. This was affirmed by the very good rating on teamwork.
3 The graduates have good potential for leadership. This was seen as a positive quality among our graduates.
4 Strong work ethics are very evident in the graduates of the university.
5 Graduates of the university can be relied upon whenever they are assigned to do a particular task.

4.4 *Weaknesses*

Despite the positive results obtained in the quantitative data collected, qualitative data showed otherwise. There is a need to address the following concerns:

1 Need to enhance skills in mathematics.
2 Need to address some students with personality problems.
3 Need to enhance the communication skills of some students.
4 Need to emphasize the value of loyalty to the company.
5 Need to emphasize their career path especially among the scholars.
6 Emphasize discipline as a virtue.

5 CONCLUSION

The present collaboration effort between the university and the industry is the scholarship program offered by the different shipping and manning companies in the Philippines.

6 RECOMMENDATIONS

6.1 *For Subject Area Heads of the University:*

1 Utilize the current data to increase the
2 performance of the students.
3 Maximize classroom supervision.
4 Implement mentoring the mentors
5 Encourage team teaching.

6.2 *For the shipping and manning companies:*

1 Give assistance in the form of donations in order for the university to acquire state of the art equipment.
2 Provide assistance in the training of the faculty in order to deliver the quality of instruction needed by the industry today.
3 Provide a career progression scheme for the graduates from cadetship to officer ship.

REFERENCES

Atlan, Taylan (1987) "Bring Together Industry and University Engineering Schools," in Getting More Out of R&D and Technology, *The Conference Board,* Research Report #904.

Peters, Lois S. and Herbert I. Fusfeld (1982) *University- Industry Research Relationships,* National Science Foundation.

Wu, Feng-Shang (1994), Technological Cooperation Model and Trend", the Third Symposium on Industrial Management, Funjung University, Taiwan.

Del Rosario (2007), *Enhancing industry-academe linkages, 2nd National Congress of the Coordinating Council for Private Education Associations*, Philippine Daily Inquirer, March 27, 2007

Santoro, Michael D. (2000), Success *breeds success: the linkage between relationship intensity and tangible outcomes in industry- university collaborative ventures.* Lehigh University: USA.

Human Factors

5. Factors of Human Resources Competitiveness in Maritime Transport

E. Barsan, F. Surugiu & C. Dragomir
Constanta Maritime University, Constanta, Romania

ABSTRACT: Studying competitiveness of human resources in maritime transport is a complex issue as it depends on several factors like organizational structure of shipping companies, social climate on board ship, multinational work environment, organizational culture, technology, safety and others. This paper tries to point out the importance of most significant factors that can be considered when discussing competitiveness in this domain and the approaches that must be taken into account for attaining a higher level of competitiveness

1 INTRODUCTION

There have been made increasing efforts to gain an awareness of human element issues. The traditional view that human error is the major cause of all accidents is being challenged by some who consider human error to be a symptom of deeper problems with the system. Errors can be induced through bad design, poor training or poor/inadequate management systems. Indeed, some argue that modern technology has reached a point where improved safety can only be achieved on the basis of a better understanding of human element within the system.

A competitive maritime company is the one that clearly states a mission reflecting a serious commitment to international transport activities and has the ability to identify and adjust rapidly to client's needs and opportunities providing high-quality, competitive navigation services. The following elements are determinants for a firm's international competitiveness: specific assets and core competences that can be exploited to their competitive advantage, reputation, continuous innovation in ship building, ship design or ship services, shown in particular in passenger cruises, a firm's architecture which describes the culture of the company.

When considering the overall picture of maritime business, personnel recruiting and training are high on the list of matters that influence competitiveness, safety and excellence in operating cargo and vessel, beside organizational structure of shipping companies, social climate on board ship, multinational work environment, organizational culture, technology, safety and others.

2 HIGHER LEVEL OF COMPETITIVENESS THROUGH RECRUITMENT AND TRAINING

2.1 *Human Resources Competitiveness through Recruitment of Seafarers*

Business environment changes quickly and there appear new needs of personnel qualifications and new ways of increasing the productivity. The growth and diversification of maritime activities has lead to an increase and an evolution of threats; this new situation requires the consideration of individual threats (navigation, accidents, terrorism, immigration, illicit traffic and pollution) and environmental threats (natural resources and disasters). In such environment, competitive seafarers are the ones who are well trained, accept a low level of risk and are responsible with their work and with the marine environment.

The traditional way of studying human performance in the maritime work domain is through the analysis of accident reports or more accurate through the analysis of accidents. 80% of maritime accidents are caused by human factors or human error. Experts who make accident reports evaluate in the first stage the human performance in the particular case against the performance standard you could expect from the crew in the given situation (Barsan et. al, 2007). According to international regulations, after any incident needing the involvement of authorities, the first thing an inspector of Marine Investigation Branch does on boarding the vessel is to check the competence and training of the seafarers on board.

A maritime company can keep a competitive advantage for medium and long term, by focusing on human resources strategies that can reach the following specific actions: the human resources orientation over client, maintaining the transparency of information needed all over the human resource department, opening and keeping new communication channels, improving communicational climate, both formal and informal, developing professional abilities and interpersonal communication skills (cooperation, improving motivation and dealing with emotions in organizational behavior, team work, etc) of the human resource. A personnel strategy in maritime transport may include a large perspective and a dynamic vision over human resources, influenced by the fact that in most cases crews are multinationals. An important element is the definition of the general objectives for medium and long term concerning human resources strategies. For establishing its own personnel strategy, a maritime company might consider increasing the efficiency of the transport services on the national and international markets to get a higher profitability, cooperation with educational institutions, crewing companies and HR companies specialized in providing professional training services. In the context of the complexity of the global labor market, recruitment of the best needed seafarers and cadets represent one of the most essential strategies for a maritime company to acquire and develop competitiveness. The recruitment process in maritime transport is the main process to bring new seafarers to the company. In today context, when a large number of sea accidents happen due to human error, a highly importance must be paid to recruitment in order to maintain quality and safety of maritime operations. Apparently, the process of recruitment and seafarers' selection seem easy, but in reality there are no easy recipes for the success of this process, as it depends on the knowledge and skills of the ones involved.

Recruitment of seafarers can be made from an existing pool of internationally trained seafarers due to expansion of the company, promotions, study leave, retirement or sickness. Historically, international shipping companies with tradition have their own „cadet" programmers to ensure a progression through ranks of seafarers who had grown up with the company's culture. This method of replacement has diminished over the last decades and nowadays companies use to outsource and collaborate with crewing companies. The benefit to the owner is a reduction in office overhead but the negative affect might be the lack of loyalty from the trainee.

Training will help improve the skills of any seafarer, but no manner of training helps if is selected the wrong person. Therefore, it is essential to carefully choose the qualities expected from the suitable candidate for the task. In every successfully maritime company there is a sense of belonging for seafarers. Instead, in poor rated companies, crews are only motivated by money.

In order to acquire competitiveness and business excellence in this business, the management of the maritime company should state a clear vision so that the crew members can understand the expectations of management. A solution for establishing derived strategies, in specific areas of the personnel activities, in consonance with the concrete conditions of the international environment is to develop partnerships for elaboration and implementation the strategies in continuous professional training.

A maritime company striving for excellence must take intro consideration that multinational teamwork, collaboration, communication and rewards for excellence contributes to developing commercial and risk management skills and provide a competitive advantage for the human resources.

2.2 *Maritime Human Resource Competitiveness through Training*

Traditionally, the purpose of training and development has been to ensure that seafarers can accomplish their jobs efficiently. Today, during the financial world crisis, the business environment has changed, with intense pressure on organizations to stay ahead of the competition through innovation and reinvention. Strategic positioning of training and development directly promotes organizational business goals and objectives.

Current trends emphasize the importance of training and intellectual capital, a critical factor for competitive advantage. The development of partnerships for knowledge sharing (e.g., consultants and/or academic partners as subject matter experts) has increased. To develop specialized training programs in corporate university settings, training departments often work closely with academic partners to prepare high-potential seafarers for leadership roles. When strategically applied, continuous learning fosters knowledge and skills acquisition to help the maritime company achieves its goals. Human resources department role is to establish and implement a high-level roadmap for strategic training and development. The starting point is an in-depth understanding of the business environment, knowledge of the organization's goals and insight regarding training and development options. HR must then develop strategic learning imperatives (high-level, learning-related actions that an organization takes to be competitive) that align with business goals. Differences in industry, business goals, human capital skills and resources influence the selection of learning imperatives (Tannenbaum, 2006).

Competitiveness can be achieved at the early stage of selecting personnel. For selection of crew members, maritime companies make a psychometric profiling, interviews, aptitude tests and portfolio determination. Entry level solutions are pre-sea courses for cadets, counseling, distance learning, STCW courses and system training. Professional competence is acquired after passing competency courses, simulation training. For professional competence, an important tools are assessment and on board mentoring. Enhancement is provided by using advanced simulation training, leadership and teambuilding programs, customized training based on the needs of the team. A constant analysis of productivity is also recommended for up to date results. The training system that a competitive company must take in consideration is not only limited to shipboard training. Shore based training implies external courses like STCW 95 mandatory course, competence and soft skills enhancement courses. In house shore based courses are related to company's system and policies, safety, productivity and loss control. The objectives of structured shipboard training are acquiring theoretical knowledge, familiarization and drills, mentoring and next rank training.

Good quality training is a prerequisite to ensuring a vessel maintains a high standard of operation. Training in all its forms adds to the value and safety culture on a vessel. From the legally imposed training certificates of competence to the cadet programs of practice at the board of the ship, it is essential to understand the strategic importance of operating a vessel to the highest levels.For crew members aspiring to higher ranks, statutory training is carried out as they seek for promotion, but a good ship operator will have their own in-house training program that will help reinforce the company's culture and safety. Undertaking responsibility ashore means learning many new skills. Skilled officers make excellent managers but the knowledge prescribed to prove competence at sea leaves gaps in a number of disciplines required in an effective competitive ship management company.

A great number of companies today have human resource sections instead of „personnel department" that were used in the past. This is the part of the company that is responsible to the CEO, the board of directors and shareholders for ensuring that the operational matters of crew deployment are carried out successfully.

A maritime company should believe in supporting employee development by using a Performance Management Plan. This is a strategic performance tool which support strategic plan and goal attainment of the organisation. The objectives of this tool is to compensate related decisions, promote the crew's potential, establish efficient rewards and recognition systems, helping objective or equitable decision making. By using a Performance Management Plan, the management can identify proper training needs (Surugiu et al., 2010).

A well-trained seafarer is the most valuable asset an owner has on board. Companies must meet the requirements for which they are directly responsible. Having well-trained seafarers is essential to any maritime company who wishes to demonstrate that responsibility, while at the same time be seen by the community as having quality and competitive operation.

3 CONCLUSIONS

Competitiveness is acquired when seafarers act as professionals in every action they make at the board of the ship or o land. Employment conditions for seafarers should be at least comparable with those found in other industries – particularly in view of the obvious impact that the quality of the shipping industry's workforce has on safety at sea and protection of the marine environment. Applying a professional way of thinking, not only in their job, but in life in general, brings satisfaction.

To maintain success in today crisis period, investment and innovation in recruitment and training programs must continue in order to achieve a competitive status.Used on a large scale, seafarers' recruitment training programs can give a competitive edge. Their aim is to select and equip new entrants to the maritime industry with the skills and knowledge to pursue a career at sea and to perform their jobs better. The training programs are also essential to gain new set of skills and knowledge for seafarers career transition.

In our opinion, competitiveness means professionalism and high productivity. At the level of maritime company, it means cost efficiency and cost effectiveness. But competitive companies do not cut costs with training systems for seafarers and invest in sending employees to international conferences and seminars on safety issues and safety management.

REFERENCES

Barsan, E. & Arsenie, P. & Pana, I. & Hanzu-Pazara, R., 2007, Analysis of Workload and Attention Factors on Human Performances of the Bridge Team, Pomorstvo (Journal of Maritime Studies), vol. 21, No. 2, Rijeka.

Năftănăilă, I. & Nistor, C., Strategic Importance Of Human Resources Recruitment And Training For Competitiveness And Business Excellence In Maritime Transport, Proc. of the 4th International Conference on Bunisess Excelence, 16-17 October 2009, 2:26-29, Brasov: Romania.

Surugiu, F. & Nistor, C. & Cojanu, G., 2010, Impact of Seafarers Training on Crew Personnel Strategy and Competitiveness, Maritime Transport & Navigation Journal, vol. 2 , No. 2, Constanta: Romania.

Tannenbaum, S., 2006, A strategic view of organizational training and learning, In K. Kraiger (Ed.), Creating, implementing and managing effective training and development: State-of-the-art lessons for practice, San Francisco: Jossey-Bass.

***, Diploma in Ship Management, Module 6, Lloyds Academy, 2004.

***, www.stcw.org

6. Human Factors as Causes for Shipboard Oil Pollution Violations

A.H. Saharuddin & A. Osnin
Faculty of Maritime Studies and Marine Science, Universiti Malaysia Terengganu

R. Balaji
Akademi Laut Malaysia (ALAM)

ABSTRACT: Shipping is a crucial transportation mode for world trade. Operation of ships has become a specialisation. Maritime training addresses the needs and in doing so is heavily regulated. STCW lays down the requirements for such training and all training patterns in the world follow these. An important aspect of the training is the environmental factor. Ships use and carry large quantities of oils. This increases the potential for pollution. The laws and penalties on this front have increased and become stricter. This has decreased the operational pollution yet, there are violations occurring. The natures of violations are not only physical but also in documentation such as falsification of Oil Record Book entries etc.

A study was undertaken to understand the effect of factors such as training, experience, attitude and fatigue on the oil pollution violations. The adequacy and effectiveness of current maritime training has been verified with reference to STCW and the recommended Lesson Plans of the IMO. Training apart, hypotheses on other human factors have been framed and tested by statistical methods. In this paper the human factors of experience, attitude and fatigue are projected and the results are discussed. The various statistical methods such as ANOVA, Chi-square and correlation analyses have been applied as appropriate to the nature of the data obtained from the survey results. The survey conducted amongst seafaring officers formed the basis for the hypotheses and the tests.

Whereas training is found to be adequate, attitude and fatigue are shown to be the primary factors affecting oil pollution violations. Negligent attitude appears to diminish with increased experience but good attitude towards pollution prevention practices remain irrespective of the variation in experience or training. The factor of fatigue has a mention in many studies and the study validates the same. The concerns on this front are highlighted and recommendations for further probing into attitude-behaviour and fatigue are suggested.

Mind-set behaviour training at management levels and pro-activeness of companies in overcoming some reasons for fatigue such as long working hours etc. are suggested. It is observed that attitude and fatigue could be the main causal factors which are resulting in pollution violations.

1 INTRODUCTION

In comparison with other modes, shipping is the most economical means of transportation and over 90% of world's trade is being routed through ships. Ships use various oils in their machineries and also transport oils. This gives scope for oil pollution of the oceans. Controlling pollution lies in the competency of the professionals operating the ships. Training provides the skills and knowledge to carry out the ship operations efficiently.

The STCW regulates the maritime training in substance as also in conduct of pre-sea training, post-sea training and assessment of competencies. The aspects of environmental pollution by oil, chemicals, garbage, sewage, emissions et al., are addressed in these standards. MARPOL (Marpol 73/78) lays down regulations for constructional and operational aspects with regards to environmental protection. Various bodies like Port State Control etc. ensure compliance of the regulations. Non compliance will result in various control measures. Towards the later half of the last century, regulations and control measures have become stricter, especially on the environmental front.

Because of these stern measures, oil spills have reduced significantly as shown in Table 1. In the recent years, violations on these types of oil pollutions have resulted in heavy monetary fines and incarceration punishments, often leading to criminalisation of the seafarer. Defences to liability are limited, as proof of the offence does not require evidence of intent or negligence (Hebden, 1995). A trend of criminalising pollution violations can be seen from the increase in the number of environmental laws like the recent European Union directive on pollution violations.

Table 1: Reduction in Oil spills (Global Scenario)
(Source: International Tanker Owners Pollution Federation Ltd., 2006)

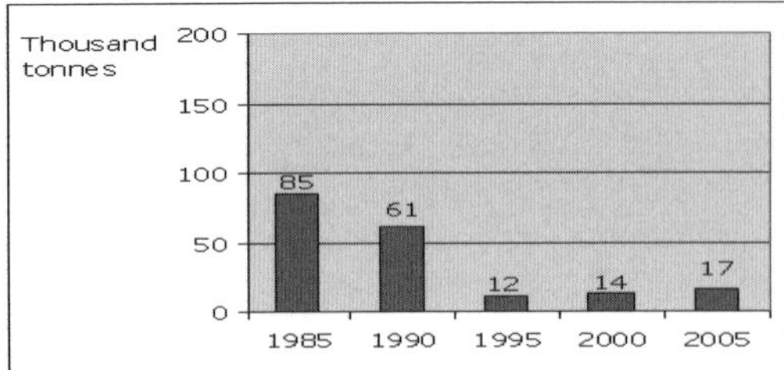

In spite of these measures, pollution violations continue to occur. BIMCO (Baltic International Maritime Council) conducted a study (2006) and Table 2 displays the number of analysed cases where sanctions were taken against the seafarers for pollution violations, after a deliberate act or negligence had been admitted or proven in court. The noticeable feature of the report is that all of them were oil pollution related offences.

Table 2: Pollution related violations; Findings in cases
(Source: BIMCO Report, March 2006)

	Pre-2000	2000	2001	02	03	04	05
Malta			1				
France							(1)
Greece							1
Singapore					1		
USA (25)		1	2	4	5	5	8
Total (29)		1	3	4	6	5	10

Note: In the French case, though the Master was fined, "it is unclear whether or not there was an intent to break the law or if negligence was involved"

Training of the officers imparts the knowledge of regulations and the serious consequences of violations. In a grosser sense, the shipboard officer is given the responsibility of shipboard functions having an environmental impact. He is termed as a "public interest" officer (Hendrik, 2006) and therefore it is recognised as a social responsibility not to violate. Yet the incidence of violations indicates that there could be other reasons for pollution violations being committed by the officer. The Paper has highlighted the results of a study undertaken under this perspective.

2 LITERATURE REVIEW

The classic Theory of reasoned action (Fishbein & Ajzen, 1975) is based on the premise that intention of a person predicts and influences the attitude of a person. Attitude towards behaviour and subjective norms are what could influence the intention itself. Attitude towards behaviour is based on what people think about the outcomes of their decision. Subjective norms are what people believe as acceptable behaviour or otherwise.

Oil pollution acts and falsification may be assumed as the intended actions (attitude-behaviour). A serious outcome will be the penalisation. With regard to norms, it is unacceptable. Based on the theory, the outcome and the norms have to be considered before any action. The action would constitute the attitude-behaviour. The most important aspect is that the seafarer must be aware of this. Awareness comes from the knowledge. Knowledge in a profession is gained from training and experience.

A traditional assumption in this regard is that increases in knowledge (knowledge quantum) are associated with greater influence of attitudes on behaviour (Fabrigar et al, 2006). A study, using an open-ended knowledge listing task, assessed attitudes toward protecting the environment. The study found that attitudes based on high amounts of knowledge were more predictive of environment-related behaviour than were attitudes based on low amounts of knowledge. It was also established with their experiments that attitudes predicted behaviour, regardless of complexity. Simply put, it may be assumed that knowledge affected the attitude of a person's profession related actions.

In an argumentative sense, with better knowledge it can be expected that a person will exhibit a better attitude while discharging his duties. Traditional knowledge dissemination formats are curriculum based training and on-the-job training. Development of operational practices is the outcome of such training. For example, if it is known that oil pollution causes harm to environment and also the person, it will lead to a concerned attitude and the work practices will follow suit. On the other hand, a person may still commit the violation being fully aware of the knowledge. This may be presumed to be the negligent attitude. In reference to context, it may be said that attitude-behaviour towards pollution prevention, therefore, might have a relationship with knowledge (training and experience). The comparative approach of the study is then justified.

Additionally, some deliberations on maritime education and training are also reviewed to substantiate the orientation of the study and its composition. Feelings and beliefs of the learner are two components of attitude identified in discussing concepts of learning. It is observed that the learner feels anxious about things he cannot do and confident about things which are achieved (Baillie, 1997). It is further observed that attitudes are closely associated with personal experiences. Knowledge, attitude and experience have a determining effect on behaviour of a person in an ambience where professional skills are put into use.

Further, a P&I Club report (UK P & I Club, 2005) on manning clearly identifies the human factors affecting the performance of the ship's staff. An analysis of the claims in Figure 1 indicates that human errors have been the causes for almost 42% of the claims.

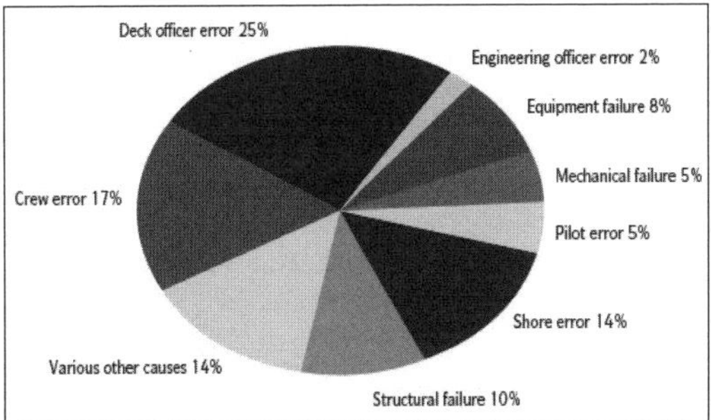

Figure 1: Main Causes of Major P&I Claims

The Report also identifies the factors which are listed in Table 3. Morale, motivation, loyalty, conditions of service and management policies can affect the attitude. It is perceived that fatigue, training and experience will have a greater influence on attitude-behaviour resulting in violations. Training may be assumed to be equal to all the officers but the intensity and effect will require verification. On the other hand, training itself can affect attitude.

Table 3: Human Factors Affecting Performance of Ship's Staff

Factors	Remarks
Fatigue	Long working hours etc.,
Morale	
Motivation	
Loyalty	
Training	Adequacy or intensity of training lacking
Language	Multinational crew
Conditions of service	
Experience	Lack of work exposure
Standards of Certification	STCW
Environment	Multi-cultural ambience
Management Policies	Companies' pro-activeness

The factor of motivation could affect the attitude towards work practices. In an analysis of human factors affecting the performance of OWS (Oily Water Separators), Hendrik (2006) makes some relevant observations. It is observed that OWS and associated systems exist for the benefit of the public rather than for the owners and the crew. From the human factors point of view, these systems are not automatically functional and additional motivational procedures are required. Two motivational procedures are mentioned, one, the threat of random and severe penalties and secondly, the incentives for whistle blowers (Hendrik, 2006). In analysing the root causes for non-compliance with pollution procedures, it is observed that behavioural causes contribute (Kumar & Loney, 2008). The most significant predictor was the 'expectation' that a procedure or a regulation must be violated because of the combined reasons of time pressure, no alternative methods and poorly constructed procedures. The other predictors are a feeling of control, opportunity for short cut and faulty planning. The study focussed with the assumption that these factors of training, attitude, experience and fatigue have affecting relationships with oil pollution violations.

The factor of training was given an exclusive treatment in the study such that the other factors were tested with relevance to training. This paper focuses on these factors alone with temporal shifts on the factor of training.

3 FRAMEWORK OF THE STUDY AND METHODOLOGY

The conceptual framework was shaped with two approaches as shown in Figure 2. The next step was to find if any relationship exists between training, the other human factors and oil pollution violations. Adequacy of training was checked first and then the examination of relationships between the human factors followed. The study relied on data obtained from a survey conducted amongst shipboard officers. The composition of the sample population was largely Malaysian but a section of Indian officers were included for better representation of the global officers. While a miniscule percentage comprised of other nationalities, in total, 522 officers were surveyed. The officer sample contained engineers predominantly as the scope for oil pollution was greater with the engineers.

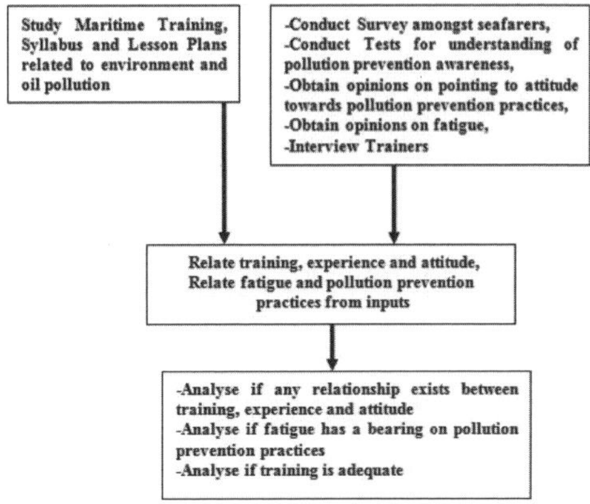

Figure 2: Skeletal Framework of the Study

Table 4. Training: Measurement Methodology

Independent Variable	Measurement Methodology	Criteria	Tested Hypotheses	Decision adequate
1.Learning content: Syllabus. IMO Lesson Plans & STCW'95 requirements	Content analysis of STCW'95, IMO Lesson Plans, training hours & Syllabi	Content & training hours = or > prescribed	Adequacy of Training	Training
2. Analyses based on test scores for groups with varying amount of training exposure (Awareness)	ANOVA	If Sig.F < 0.05 then Reject H_{O1}	H_{O1} & H_{A1}	Since Sig.F > 0.05 then Accept H_{O1}
3. Analyses based on acceptance and non- acceptance to violations for groups with varying amount of training exposure (Attitude)	ANOVA	If Sig.F < 0.05 then Reject H_{O2}	H_{O2} & H_{A2}	Since Sig.F > 0.05 then Accept H_{O2}
4. Analyses based on acceptance to involvements in pollution violations for groups with varying amount of training exposure	CHI SQUARE	If Sig.$\Psi^2 < 0.05$ then Reject H_{O3}	H_{O3} & H_{A3}	Since Sig.$\Psi^2 > 0.05$ then Accept H_{O3}

Table 5: Human Factors: Measurement Methodology

Independent Variable	Measurement Methodology	Criteria	Tested Hypotheses	Decision adequate
1. Analyses based on test scores for groups with varying amount of experience (Awareness)	ANOVA	If Sig.F < 0.05 then Reject H_{O4}	H_{O4} & H_{A4}	Since Sig.F > 0.05 then Accept H_{O4}
2. Analysis based on acceptance and non-acceptance to violations for groups with varying amount of experience (Attitude)	ANOVA	If Sig.F < 0.05 then Reject H_{O5}	H_{O5} & H_{A5}	Since Sig.F > 0.05 then Accept H_{O5}
3. Analyses based on correlation between citing fatigue as a reason to acceptance and non-acceptance for violations	Correlation Analysis	If Spearman's Coefficient, $\rho < 0.05$, then Reject H_{O6}	H_{O6} & H_{A6}	Since Spearman's Coefficient $\rho > 0.05$ then Accept H_{O6}

Table 6: Summary of Results-Hypotheses

Hypotheses	Result
Training could be inadequate	Training is adequate
H_{O1}: There is no significant difference in levels of oil pollution prevention awareness between officers with varied hours of training H_{A1}: There is significant difference in levels of oil pollution prevention awareness between officers with varied hours of training	Accept H_{O1}
H_{O2}: There is no significant difference in attitude towards pollution prevention practices between officers with varied hours of training H_{A2}: There is significant difference in attitude towards pollution prevention practices between officers with varied hours of training	Reject H_{O2}
H_{O3}: There is no significant relationship between number of hours of training and involvement in oil pollution violation incidents H_{A3}: There is significant relationship between number of hours of training and involvement in oil pollution violation incidents	Accept H_{O3}
H_{O4}: There is no significant difference in levels of oil pollution prevention awareness between officers with varied years of experience H_{A4}: There is significant difference in levels of oil pollution prevention awareness between officers with varied years of experience	Reject H_{O4}
H_{O5}: There is no significant difference in attitude towards pollution prevention practices between officers with varied years of experience H_{A5}: There is significant difference in attitude towards pollution prevention practices between officers with varied years of experience	Reject H_{O5}
H_{O6}: There is no significant relationship between fatigue and pollution prevention practices H_{A6}: There is significant relationship between fatigue and pollution prevention practices	Reject H_{O6}

Six hypotheses were framed of which, 5 were based on the human factors of attitude, experience and fatigue. The adequacy of training was checked by content analysis of maritime training (engineering stream) syllabi with reference to STCW (Standards of Training, Certification and Watchkeeping) and IMO Lesson Plans. The statistical tests were chosen according to the nature of the data and the type of hypothesis. Appropriate criteria were established for acceptance or rejection of the hypotheses. The tests and criteria for validating the assumptions of the study are projected in Table 4 and Table 5. Further inputs were obtained from trainers attached to maritime institutes.

4 RESULTS AND DISCUSSION

The content analysis of the training syllabi and IMO Lesson Plans showed no apparent lack of training. The quantitative training appears to be sufficient with 17 hours, which is well above the 15 hours indicated in the IMO Lesson Plans in the post-sea scenario. The hours of exposure to training on pollution get enhanced if pre-sea quantum and the modular courses were also considered. The other hypotheses were verified by statistical tests, the results of which are summarised in Table 6.

Tests of bad (negligent) attitude and good attitude were conducted. While a deviation from normal, legal oil pollution prevention practice was considered as a bad attitude, conformance to rules was considered as good attitude. The sample population was divided into 5 groups based on the levels of training exposure and it was assumed that attitude differed with the training exposure. Based on the ANOVA, the next hypothesis, H_{O2} is rejected with Sig. levels being equal (0.05). Also, the post-hoc tests indicate a decline of test scores for bad attitude with an increase in number of training hours. However, tests on good attitude showed no difference (Sig.0.611>0.05) and post-hoc tests showed no variation in average scores. It may be well assumed that existent good attitude does not enhance or diminish with increase in training hours. Good attitude is prevalent irrespective of the amount of training, while bad attitude reduces with increased amount of training. A parallel may be drawn with the attitude-behaviour patterns being affected by knowledge (Fabrigar et al, 2006). Increase in knowledge (training) does influence the attitude-behaviour.

The next hypothesis to be tested was to see if training made a difference to the officers' involvement in pollution incidents. A similar test was done for the human factor of experience assuming that increased experience will reduce involvement. Assuming increased training would mean increased experience, the involvements were tested with groups with varying experience. With Sig. $\Psi^2 = 1.00 > 0.05$, it is seen that increase in experience does not affect (bring down) pollution violations. Though this validates the acceptance of H_{O3} where training was the factor, the relationship with experience was not tested by framing a hypothesis. It was seen that a hypothesis relating experience and involvements might fall into Type I error (Rejection of the Null when True). This is because an inference showing that increased experience increases number of violations might not be true, as the scope for pollution violation increases with increase in experience (work period).

The next test was on experience and awareness. Training on pollution matters is enhanced even after the shipboard officer reaches high ranks. An increase in experience exposes the officer to increased training hours and hence the knowledge. Officer sample was grouped into 7 varying levels of experience and it was assumed that experience would increase knowledge of pollution awareness. Results of H_{O4} confirm this. With Sig. 0.576 > 0.05 but HOV Sig. 0.013 < 0.05, H_{O4} is rejected, leading to the inference that increase in experience increases awareness. This outcome is not similar to that of H_{O1}, where groups of varying training hours were tested for awareness. This syllogism leads to a conclusion that quantitative training has the same intensity to all the officers who undergo training, whereas, with gain in experience the quality of the training gained (knowledge) improves.

Experience and attitude measures were tested next. Proceeding to next hypothesis, with Sig. 0.187 > 0.05, but HOV 0.037 < 0.05, H_{O5} is rejected. The scores for bad attitude tests show a decline with increase in years of experience. Increase in experience is seen to diminish bad attitude towards pollution practices. Further tests of experience with good attitude test scores showed no relationship (Sig.0.157 > 0.05) between them.

This is similar to the outcome of H_{O2}, where increased training diminished bad attitude but did not affect good attitude. It may be inferred that increased experience and training hours lessens the bad (negligent) attitude towards pollution prevention practices though existent good attitude does not enhance or diminish with increased training and experience.

The last of the hypothesis, H_{O6} was tested and the factor of fatigue showed significant relationship towards pollution prevention practices at $\alpha = 0.01$ itself. With this, H_{O6} is rejected. Here, it was assumed that fatigue affected the shipboard operational practices particularly those pertaining to oil pollution prevention. This is further supported by the survey opinion shown in Figure 3, where a maximum number of respondents have identified fatigue as the major factor causing difficulties in MARPOL (pollution prevention) practices. Also, fatigue is the primary factor affecting performance of shipboard staff as identified in the P&I Report on manning (2005). The

IMO guidelines on fatigue (2001) and measures for mitigation prevail with the assumption that fatigue is a major factor affecting performance. The results of the study validate the same.

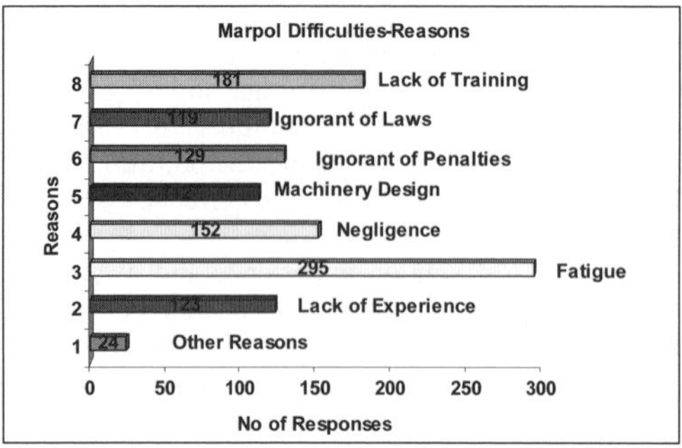

Figure 3: Reasons for Facing Difficulties in MARPOL Practices

Other tests of correlationship between experience, awareness test scores and attitude test scores showed no significant relationship. Amongst these tests, good attitude test scores comparatively showed a higher relationship with experience. But ANOVA scores were given more credibility as ANOVA tests are statistically more reliable for data being scalar.

5 CONCLUSION

The study recommended realistic approaches to enhance training by development of simulator exercises, increasing case studies, upgradation of the trainers' knowledge, treatment of environmental protection as an independent subject etc. Focussing on the other human factors, the study highlights few issues of concern. The recommendations are apparently based on the over-all results summarised in Figure 4.

The basic factor of training may be assumed to have been imparted effectively. Experience, on the other hand is a factor where strategic control is not possible. Attitude-behaviour is a major causal factor as much is fatigue considering the oil pollution violations. In the first place, further studies in these areas must be undertaken. Many factors may be cited which could be affecting attitude-behaviour and fatigue. Multi-cultural ambience, long working hours, absence of mate, extended shipboard stays, pressure from superiors and principals etc. are a few worth a mention. With projected shortages in ship manning, these issues will adversely contribute to the deviant behaviour causing pollution violations.

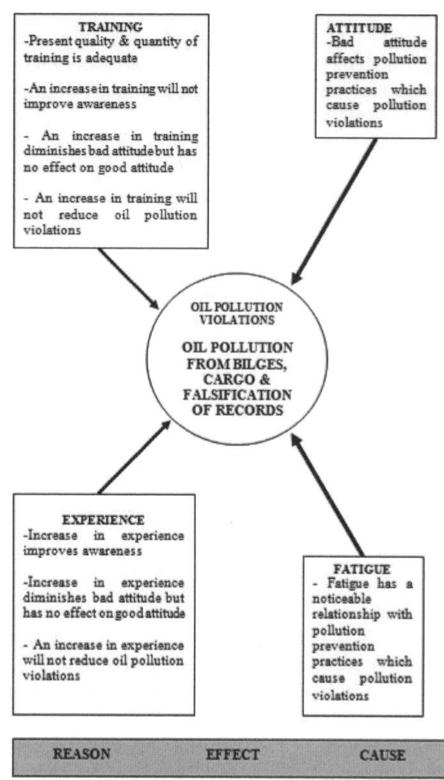

Figure 4: Effect of Training and Human Factors on Oil Pollution Violations

As a paradigm approach, training on mindset-behaviour at early stages of an officer's career (less experience) may be included in the training formats. Many maritime training programmes around the globe follow regimentation systems. The systems apart from preparing the officers for the hardships of sea life also aim to inculcate ethical behaviour.

With regards to fatigue, the industry must find some creative solutions. It is observed that because of fatigue, reaction times when faced with a problem are greater and since concentration is also affected, the probability of an accident increases (Gonzales, 2000). Rearrangement of watch hours ensuring proper rest periods are being tried out which differ from the traditional four on-eight off patterns. A Finnish study in fatigue (2008) identifies specific factors such as sleep apnea, lack of fresh air and time of the day in the watch-keeping schemes as some of the factors influencing fatigue. Timely relief and healthy work ambience are other issues for shipping companies to be pro-active about. In the broader scheme of things, attitude of the shipboard officer and the overbearing effect of fatigue might remain as ever present issues for the industry to contend with.

REFERENCES

Baillie, Don, (1997). Section I, Chapter 2, Concepts of Learning and their Application, Maritime Education and Training – A Practical Guide, 1997, London, The Nautical Institute in conjunction with World Maritime University, ISBN 1 87 00 77 415, pp.10, 11 & Section I, Chapter 3, Concepts, Skills and Competence in a Maritime Setting, pp.18, 20.

BIMCO (10 Mar 2006). BIMCO Study of recent cases involving the International Practice of using Criminal sanctions towards seafarers, (Adopted by BIMCO Board of Directors on 2 Mar 2006, Rev. 10 Mar 2006) & The Presentation of the Study (14 Mar 2006) at IMO/ILO ad hoc Expert working Group on fair treatment of seafarers.

Fabrigar, L. R, Petty, R. E, Smith Steven, M & Crites, S. L. (2006). Understanding Knowledge Effects on Attitude - Behavior Consistency: The Role of Relevance, Complexity, and Amount of Knowledge, Journal of Personality and Social Psychology, Vol. 90, No. 4, pp.556–577.

Fishbein, M, Ajzen, I. (1975). Belief, attitude, intention, and behavior: An introduction to theory and research. Reading, M A: Addison-Wesley.

Gonzalez, Blanco, (2000). Section 3, Analysis of Pollution Incidents at sea and those caused by port operations, Maritime Engineering and Ports II, 2000, WIT Press, UK, ISBN 1-85312-829-5, pp.165-168.

Guidance on Fatigue Mitigation and Management, (12 June 2001). IMO MSC / Circ.1014, Ref T2/4.2.

Hebden, D. G., & Sheehan, C. (24-25 May 1995). The duty of ship owners with regard to safety and pollution prevention, Paper 4, IMAS 95, IMarE Conference on Management and Operation of Ships: Practical Techniques for Today and Tomorrow, Volume 107, 2, pp. 57.

Hendrik van Hammen, F. (January, 2006). Initial Recommendations for Bilge Oily Water Separator System Design and Operation, MEETS Symposium, Arlington, USA, Reprinted in Marine Engineers Review, February, 2006, pp.26, 24.

Investigation Report, (2008). Factors contributing to fatigue and its frequency in Bridge work, Accident Investigation Boards, Finland, Translation of the Original Finnish work, ISBN 951-836-225-4, pgs 44-45.

Kumar, Suboth & Loney, J.S, (August 2008). Marine Environmental Excellence, Route from Compliance to Excellence – A sustainable way, Marine Engineers Review (India), pp. 25.

Marpol 73/78 International Convention for the Prevention of Pollution from Ships, 1973. As modified by the Protocol of 1978 relating thereto, 17 February 1978, A.T.S.1988 No.29.

The Human Factor: A Report on Manning, UK P&I Club 2005.

STCW Convention

7. Needs and Importance of Master Studies for Navigators in XXI Century and Connectivity to STCW 78/95

A. Alop
Estonian Maritime Academy, Tallinn, Estonia

ABSTRACT: The high-level technology and equipment of modern ships from one side and the constant growth of high stress and psychological pressure to ship officers, especially on management level, from other side pose a number of important questions relating to optimal combination of professional knowledge and skills and personal characteristics of seafarers. What may be the best combination of different subjects and courses in training programmes? What may be the role and place of master studies in evolving of high qualified specialists from one side and strong and self-confident personality from other side? What should be the most important difference of master study programmes for ship officers from those for land-based personal? The author of this presentation tries to discuss these questions relying to his long-time experience in field of maritime education and training and in organisation and carrying out of master studies in maritime academy.

1 HYPOTHESIZING OF PROBLEM

Taking under investigation the main problems and bottlenecks in field of providing of navigation safety during last 20-30 years, we may detect the paradox situation: the rate of accidents at sea doesn't show a tendency for decreasing, so that annual ratio module of ships totally lost related to number of ships in whole world merchant fleet is today more or less the same as 20-30 years ago. By first look, it seems to be at least strange because the technical and technological level of nowadays ships is much higher than some decades ago. The devices, appliances and apparatus that complete the modern ship bridge and engine rooms are in general relatively easy-going in exploitation and user-friendly. The professional preparation of seafarers is regulated by world-wide acknowledged international regulations, first at all by STCW 1978 Convention. The majority of maritime schools and academies have consistently been evaluated by competent authorities and their programmes and courses are recognized as meeting the requirements. But why we have no today the significant improving in statistics of marine accidents in whole or, at least, speed and scope of dynamic of their decrease don't satisfy us?

Author of present paper has disserted subject in question more than once, for example by presentation on the 13th International Conference on Maritime Education and Training IMLA 13 "Safety at Sea through Quality Assurance in MET Institutions. Quality Assurance in Action". In paper presented on conference author tried to carry through some initial analyses and to bring forward the point of view that the main factors bringing on the high accident rate at sea are not the technical once but rather lay in sphere of human nature and personality (Alop, 2004).

Some developments during period from 2004 up to date affected author's positions and viewpoints on matters under discussion. First of such developments is "The Manila amendments to the STCW Convention and Code" adopted at a Diplomatic Conference in Manila, the Philippines in June 2010 and set to enter into force on 1 January 2012 (www.imo.org). According to amendments there will be changes to each chapter of Convention and Code, among these new requirements for marine environment awareness training and training in leadership and teamwork. This fact shows growth of understanding of decision-makers on the highest level that promotion of technical knowledge and skills of seafarers is not enough for breakthrough in problem of improving of marine safety but so-called "soft values" will play significant or may be even crucial role in achieving of that. But, at the same time, the author's numerous discussions with seafarers and instructors in MET institutions allow supposing that it's not very clear what kind of teaching and training methods as well as teaching materials might provide the effective results in training of leadership and teamwork of seafarers. And how to measure and evaluate properly results of that?

Secondly, the curriculum for master studies with speciality for navigators has worked out and launched since 2006 in Estonian Maritime Academy (EMA). Author of this paper was directly involved in process of development and improving of this curriculum during all the last 5 years and is the Head of Master Studies in EMA at the present time. The experience and knowledge in field accumulated during this period afford him to make a presumption that master studies are very suitable stage of training for development of personal characteristics of seafarers. How and by what schemes of organization of training it is achievable in the best way is the question for what author tries to find some answers hereafter.

2 THE ANALYSES OF TRAINING SCHEMES

2.1 *Training triad today*

It's clear that ship officers, first at all navigators are the key persons in providing of safety on board. High level and good quality of their professional knowledge and skills, good seafaring experience and great personality play a crucial role in achieving of high level safety of vessel, people and environment at sea.

Let's examine this triad [i) knowledge and skills, ii) experience and iii) personality] from assimilation and trainability point of view. Obtaining of all the necessary for seafarers professional knowledge and skills is more or less wholly the result of well organized training (incl. training on board), simulation and assessment processes. The STCW 78/95 Convention gives by its Code enough good and effective instruments for achieving of professional training goals erected. The personal characteristics of students have importance on this stage only as assisting means making training process easier or, in opposite, more difficult for each person. More sluggish brain just need more time for understanding and adopting of lessons; indecision and undiscerning have not crucial importance for passing of occupational tests and exams although may lead to poorer grade.

Speaking about seagoing experience it's clear that experience may not be obtained faster by some accelerated training lessons. The figure of speech "to share experience" seems to be a little wrong and actually is misleading because experience is not professional knowledge that really may be shared. Experience is extremely personal and very valuable exclusive kind of knowledge and skills that may be obtained only by going through real life situations. In this case the personal characteristics play much more important role and they determine what the "life lessons" will be given to actor and how useful this lessons will be for him in possible substandard situations in future. Question is: how is it possible to obtain an experience of optimal acting in critical situations by training? Whether is it possible at all?

Third "pillar" personality is not something spontaneous but it is tightly connected to all processes incl. training and especially working life. As was showed above, the personal characteristics such as discretion and decision-making ability, self-determined intellectual power, presence of mind and courage to take upon responsibility or to hand it out to other persons are not really important in process of traditional professional training, carried out in big part applying the "teaching methods" "do like me" or "do according to instructions". Unfortunately, this approach is favoured up to present day by the STCW 1978 Convention and its Code. Adoption of "The Manila amendments to the STCW Convention and Code" gives us the hope for changing these attitudes in the near future.

The personal characteristics are much more important in process of obtaining of seagoing experience. What higher position in ship hierarchy than more important to have the good certain personal characteristics. For officer in charge of a navigation watch the ability for good teamwork is the most important, for chief mate and especially for captain it is important to be a leader and to have ability to manage and direct the people. Actually, all these qualities and mutual relationships manifest themselves mainly during job life on board. Trainees on board may keep an eye on these developments but they may not be real actors in them.

Having of great personal characteristic like listed above or lack of them gains the crucial importance in critical situations. In fact, a lot of human lives and huge material values may be saved or lost depending of them. Are the great personalities always inborn or they may be partly or even in whole obtained? Is it possible to acquire them only by method of "experiments and errors" in real life or there are some methods for training them?

It is complicated to give some definite and complete answers to these questions. The STCW Code foresees two courses what are intended to certain extent for development of personal characteristics. First is Bridge Resource Management course (BRM), this is also called Bridge Teamwork Management. Normally it is a three day course of instruction. Second one is Crowd and Crisis Management (CCM) and it may be one to four days long depending on training organization.

Unfortunately, in author opinion, these courses giving in usual learning environment of maritime schools and academies can't solve to needful extent problem of training out the students' personality and obtaining of essential personal characteristic in process of curriculum studies, i.e. in maritime schools and academies. The students don't have some significant sea-going experience or their experience is limited to trainee's seagoing training on-board. Abovementioned courses, as rule, are theoretical for them and come by well known method "do like me" or even "think like me".

2.2 *The Bologna process implementation*

The aforesaid is related to so-called one-step training system for deck officers, being in use in EMA during last two decades. EMA as educational institution may be identified as so-called Professional Higher Education Institution (according to Estonian education system) or as the University of Applied Sciences. Students of navigation faculty obtain theoretical knowledge and skills according to STCW 1978 requirements for both operational and management level during four years and they have additionally one year sea-going training on-board what officially is not included into duration of curriculum. These studies allow obtaining the qualification of officer in charge of a navigation watch after graduating on bachelor level. Formally, for achieving of captain position the graduates don't need coming back to school for additional training (of course, except the obligatory refreshing courses envisaged by STCW). It is only the matter of their seagoing career and practical experience. But, in author opinion, they, having good professional knowledge and skills at the end of their studies in school, have not enough good preparation in field of teamwork, leadership, team management, acting in substandard and critical situations and so on.

In year 2006 the 1,5 year long curriculum of master studies was worked out and implemented in EMA. The name of curriculum is Maritime Studies and one of three specialisations is Ship Maintenance and Navigation (SMN). In fact, this is a joint curriculum of EMA and Tallinn University of Technology (TUT) and it is oriented to giving to students wider knowledge in so-called academic subjects (subjects of TUT) as well as in professional subjects on higher than bachelor level (EMA subjects).

List of main EMA and TUT subjects see Table 1.

Regarding to Bologna system this is 4+1.5 (330 ECTS) long higher education studies for obtaining of master degree (one year of sea-going on-board training on bachelor level is not accountable for academic duration of curriculum).

All the students of master studies in EMA are working people. This is why a big part of learning is a distance learning and contact hours take place in the evening time. The most part of SMN speciality students are active seafarers (both navigators and engineers). Despite to that the curriculum is highly popular amongst graduates of EMA.

Table 1. Some subjects of master studies programme Maritime Studies

Subjects of EMA	Subjects of TUT
1 Research methodology	1 Foreign language for science and research
2 International public maritime law	2 Financial management
3 Risk management in shipping	3 Introduction to information systems
4 Hydrodynamics and seagoing characteristics of vessels	4 Investment analysis
5 Optimization of navigation	5 Quality and productivity management
6 Shipping company management	6 Project management
7 Safety and security management in shipping	
8 Merchant shipping law	
9 Ship chartering and agency	
10 Organizing of work and shipping economics	
11 Environmental pollution prevention and pollution control	
12 Automated control systems of ship	
13 Ship design and architecture	
14 Navigation safety control systems	

Looking at list of subjects in Table 1 one can see that students whether deepen and expand their occupational knowledge (EMA subjects 3, 4, 5, 8, 12, 13, 14) or prepare for themselves "springboard" for jump into future onshore working life (EMA subjects 1, 2, 6, 7, 9, 10 and all the TUT subjects). Taking into account the fact that for majority of them the main part of their seagoing career still ahead, it seems to be reasonable to bring to studies more subjects and courses that may assist them in developing of not only occupational competence but also the personal qualities essential for solving of complicated situations arising in management and administration of ship or in critical situations. The question is not only what courses and subjects should they be but also how to build them up and conduct them in such way and by such methods that result will be the most productive and efficiency?

3 THE COMPLEXITY APPROACH

3.1 *Tests for entrants*

First at all it seems to be essential to carry out the psychological test for youngsters wishing to entrance to maritime schools for navigation studies. Although the number of personal characteristics may be developed during studies and working life period, there are some of them what may be only congenital. The duty of such psychological test would be to check out persons who are improper for seagoing career by reason of their personality. In this case we have to deal with so to say total career-unfitness.

The ratio module of such cases that may be revealed on the very early stage with high extent of certitude is definitely not big, probably less or even more less than 1%, but it seems to be very important to detect such youngsters before entrance to school because in course of professional studies an exposure of such problems is not real. There are known the cases when students were enough successful in their theoretical studies and hadn't any problems before going to first long-time seagoing on-board training. But after some weeks on board of vessel they had a serious depression and other psychological problems and were evacuated from ship before end of practice and had to say good-by to seagoing career at all.

3.2 *"Sandwich" type studies?*

From point of view of development and assurance of personality of seafarers the two-step or so-called "sandwich" training system for navigators seems to be more appropriate and effective that one-step or continuous system.

This last mentioned system is applied now in majority of MET institutions and it has undoubtedly a lot of advantages. For instance, training institutions may build up the learning process in the most optimal way economically and methodically giving continuously students the professional knowledge and skills (at least theoretical part) on both operational and management level (Põldma, 2010). Problems with bringing people back to school for management level studies are not to take place as well as seafarers problems with taking up such studies. However, students having some seagoing experience by on-board training during studies have no to good an extent experience in management and administrative job on board as well as in human relationship and teamwork. It seems to be quite ineffective to teach theoretically leadership to people who didn't have any possibility to try to be leader in real situations.

In author opinion, the "sandwich" system seems to be more effective for achieving of good results in leadership and teamwork as well as in development of personal characteristic that essential for successful acting as team leader and strong person.

Taking as example the training system of the EMA it may look as following. Period of training on bachelor level should be shortened up to three years (180 ECTS) of theoretical studies plus one year of seagoing practice and this will cover only training for operational level. Professional output will be as before the officer in charge of a navigation watch having the bachelor academic degree. Studies can be made one year shorter thinking to transfer of some management level subjects and courses to master studies stage and to reducing of capacity of so-called academic subjects to minimum (the criterion is that graduates obtain the academic grade as bachelor according to provisions of education system). The stress of first stage studies should be placed on good professional preparation.

The second stage of studies will be the master studies with two years duration. The master studies will contain some high level professional training needed for essential competence on management level and a lot of academic subjects. It's necessary to work out and implement to curriculum the block of "personality studies" that will allow developing properly personality of future captains and chief mates. But this block shouldn't consist of theoretical subjects only and teaching tools and methods applying for these studies should be something different taking into account specificity of audience.

3.3 *Teaching methods and tools*

The most advantage of students' audience on stage of master studies from point of view of psychological and personality training is their seagoing experience. This includes undoubtedly to less or more extent experience in ship management and administration and, of course, good teamwork experience. It's very probably that some of students got together with their ship into more or less difficult or even critical situations, so they have such kind of experience as well.

Because students will have very different job positions on their ships and very different seagoing experience, it will possible to apply for "personality studies" some teaching methods what are the most suitable for specific audience of experienced seafarers.

Author of this paper have had lucky to take in period of 2004-2008 part in projects "Securitas Mare" and "Securitas Mare II", main task of which was working out and implementation of CCM course for seafarers on European level (www.tg4transparency.com). Such 3.5 days long course was successfully worked out, tested by number of trial courses in different European countries and got a full approval from DNV. In author opinion, the most important and valuable result of these projects was the successful approbation of the advanced teaching and learning methods used in trial and demo courses. As the main goal of CCM course is to learn how the people crowd may be restrained in critical situations, this course is very strongly about personality and personal characteristics of seafarers, especially on top-management level as well as deal with psychological problems, leadership, and teamwork and so on. So, all the know-how and experience obtained during running of these projects are very suitable for applying in abovementioned block of "personality studies".

Let's name these methods what, in author opinion, may be successfully used in organisation of "personality studies" for active seafarers.

Firstly, this is so-called "dual-instructor" method. The core of this method is that the course is carried out from beginning to end by two instructors, who work in tandem and run course in regime of dialog with students and one with another. They have to have a different background and very good experience (to be experts in field). In our case one has to be well experienced captain and other high-level psychologist. The professionalism of such persons as teachers, and their personality and ability to work together has a crucial importance for success of studies.

Secondly, it's reasonable to take in use experience based learning (EBL) method. This method based mainly on skilful using by instructors the cases bringing by students and their experience. In fact, the students enhance each another by presentation and discussion of situations they had in their seagoing career so far. The main task of instructors is to mould these discussions and to solve didactic problems by comments from different points of view and making of conclusions. Effective using of this method is possible only if audience consists of persons who have more or less experience in field of studies. As it was shown above students of master studies will have as rule to certain extent such experience, so EBL method should be suitable for "personality studies".

Thirdly, the method of learning by doing (LBD) shall be used maximally. Lectures are not typical for such type of studies; the main learning tools are discussions, practical exercises, group works. As in abovementioned CCM course, the very importance instrument for achieving of course goals is a practical drill on board of real vessel, in course of what the students will get into complicated and unexpected situations and will be obliged to find optimal solutions.

Courses curried out in such way must have the maximum effect for development and training of essential personal characteristics of seafarers. Furthermore, this block of "personality studies" may be offered successfully not only for students of master studies but be one of refreshing courses for active seafarers in whole.

By author experience, the crucial factor for success of such "personality studies" is the quality of abovementioned tandem of instructors. Their professional and personal background is so different that the creation of such high professional level tandems is not very easy. There may be some other problems with approving and financing of such schemes by authorities of maritime schools, so it may be reasonable for MET institutions to find some common solutions.

CONCLUSIONS

The personality of seafarers is not less important than professional knowledge and skills. Moreover, in critical situations the personal characteristics of captain and/or other decision-makers on-board may become even more important taking into account the crucial importance of right and effective acting for saving of human lives and material values.

The using of traditional teaching and training methods may supposed to be enough good for achieving of professional training goals. As according to "The Manila amendments to the STCW Convention and Code" these methods are at least insufficient and shall be overestimated and more effective and non-traditional methods should be found out.

It is necessary to discuss and find out the appropriable answers for important question: how to build up the system of MET for seafarers providing the achieving of the most effective results not only in professional preparation and training but in development of the great personality as well.

REFERENCES

Alop, A. 2004. Education and Training or Training contra Education? Proceedings. 5-12. *Safety at Sea through Quality Assurance in MET Institutions. Quality Assurance in Action. 13th International Conference on Maritime Education and Training IMLA 13. St. Petersburg, 14-17 September 2004.*

http://www.imo.org/MediaCentre/PressBriefings/Pages/STCW-revised-adopted.aspx, ch. 09.03.2011

Põldma, K. 2010. Merendusvaldkonna magistriõppe õppekavade arendamisest ja kvaliteedi tagamisest. Master Thesis, unpublished.

http://www.tg4transparency.com/Events_files/Securitas%20Mare.pdf, ch. 09.03.2011

8. Implementation of the 1995 STCW Convention in Constanta Maritime University

L. C. Stan
Constanta Maritime University, Constanta, Romania

ABSTRACT: The development of the maritime transportation and its connected activities imposed the necessity of having more trained people involved in this operation, able to act in very various situations based on a considerable volume of knowledge. To achieve these standards, the training process, especially for operation, safety and security activities, must be highly professional and in concordance with the international requirements in the field. This professional training involves the use of the latest developed techniques, as simulators, and dedicated computerized programs. Inside of this concept of uniform and updated training, Constanta Maritime University developed its profile and vision as maritime officers formative, combining the STCW 95 Convention requirements with the latest technological development in order to provide to the international shipping market better trained people, capable to apply STCW objectives and also to use the modern technology for a more safety sea.

1 INTRODUCTION

Today, Constanta Maritime University is the principal academic training institution in Romania. This position was acquired through a continuous effort to offer to the future deck and engine officers the best training and knowledge in the interest field. In this respect, changes were made, starting with revaluation of curricula, brought it more closely to the present requirements of the STCW Convention and shipping industry, succeeded by the improvements of teaching methods, usage of the high technology and newest simulators in this process, and, finally, but not the last, improvement and increase the level of the trainers and teachers accordingly with the latest technological development in this area of training.

The development of the maritime transportation and its connected activities imposed the necessity of having more trained people involved in this operation, able to act in very various situations based on a considerable volume of knowledge.

To achieve these standards, the training process, especially for operation, safety and security activities, must be highly professional and in concordance with the international requirements in the field. This professional training involves the use of the latest developed techniques, as simulators, and dedicated computerized programs. These new techniques and working procedures with a view to a better skills development represented, at beginning, a challenge for the traditional maritime academic training field, some of them still being a challenge due their continuous improvement and updates.

Once the shipping industry grew up the work force market requested more professionals and specialized persons, the training system, at all levels, but special at academic level, had to accept the challenge of necessary technology in order to respond and provide required personnel.

In all cases, where it was necessary to change the traditional way of teaching and practice to the new one, the first step was represented by the mentality changes of the trainers involved and, in the same time, by the update of the theoretical base, including the technique aspects. This was not an easy process, the beginning and first stages were complicated, partially due to the reduced knowledge on the new technologies and the better approach way to perform the best training in order to reach the proposed results. These difficulties were not finished once the familiarization started, they continued after this stage because the technological changes appeared soon with new products and also new procedures.

2 THE ROLE AND APPLICATION OF TE STCW CONVENTION IN THE SEAFARER TRAINING

Despite its broad global acceptance, it was realized in the late eighties that the STCW Convention did not achieve its initial purpose. Instead of it, the Convention gradually lost credibility as its widened acceptance. The main cause of this situation was the general lack of precision in its standards, the interpretation that was left to the satisfaction of the Administration. This resulted in a widely varying interpretation of its standards.

Also, at the aforementioned comment, another idea must be added that many changes had taken place in the structure of the world merchant fleet and in the management and manning of ships, since the development of the Convention in the seventies.

Thus, the necessity of an update of the Convention is required. During the 1995 revision process, no changes were proposed to the articles of the Convention so as to allow the amendments to be adopted and enter into force by means of the tacit acceptance procedure. This procedure can only be applied to the amendments made to the annex of the Convention.

Over the years passed since the 1995 revision, parties have applied for revisions and amendments to the convention. IMO's specialist sub-committee on Standards of Training and Watchkeeping (STW) prepared a draft convention accordingly that was discussed in detail and signed in Manila, Phillipines in June 2010.

Taking into consideration the revision of the STCW Convention and Code was made long after its appearance, moment when basic concepts were considered outdated by the actual requests onboard the ships, the reappraisal of the IMO Model Course in order to cover present lacks especially for the use of the modern technologies in maritime activities was also necessary. The STCW Convention provides the requirements necessary for a trained person involved in the operations on sea and the IMO Model Courses explain how can be satisfied these requirements and what areas have to be covered during the training process at the higher educational level for an operational or managerial officer.

During all these revisions, general or specific provisions were taken, updated in this way the convention and IMO model courses. There were stipulated, inter alia, ideas on the continued professional competence, guidance rules for the use of the distance learning and e-learning procedures and for assessing the trainee's progress and achievements, considering that both deck and engineer officers at management level can have a certain proportion of the required courses under distance education/e-learning schemes while they serve on board. Thus, it shall be possible to reduce their time spent at shore-based training institutions. Distance education/e-learning schemes have also to be used for lifelong learning and refreshment trainings of seafarers. In the same area of interest, one issue regarding to distance education/e-learning was the establishment of a procedure to prevent hacking and fraud of the educational scheme and therefore safeguarding the copyrights of the educators and the institutions.

The simulator training is foreseen where appropriate; according to the revised STCW Convention, the simulators must be used more effective in the training process of the future seamen and officers.

Requirements about the use of computerized techniques and specialized software for applications in the coastal and celestial navigation are necessary to be introduced in the courses model.

The high technology has to be used in order to increase the level of training and to reach higher standards of knowledge and skills. This conclusion comes from the actual situation on board ships, where the higher techniques are already present.

The responsibility of the maritime universities, as education and training institutions, is to respect in their curricula the indications given by STCW - IMO documents and to do the training accordingly.

The maritime university has the role to provide to the world fleet officers that are more trained and capable to work with latest equipment. It is unreasonable, in a century of high techniques, to teach about procedures in a field as navigation, but not to mention about the latest technology found onboard, made just to be used for a more safely navigation activity.

In order to improve competences and skills of the future officers, it is important to make changes on the present training requirements and bring them to the actual development of the maritime field and so, to consider the missing is covered.

Staying in the same subject area, at this moment, the training of the future maritime officers at Constanta Maritime University is made under the recommendations of the International Maritime Organization, according with the STCW Convention and IMO Model Courses for operational and managerial levels

Inside of this concept of uniform and updated training, Constanta Maritime University developed its profile and vision as maritime officers formative, combining STCW 95 Convention requirements with the latest technological development in order to provide to the international shipping market better trained people, capable to apply STCW objectives and also to use the modern technology for a more safety sea.

3 THE DEVELOPMENT OF THE TRAINING PROCESS IN CONSTANTA MARITIME UNIVERSITY ACCORDING WITH THE STCW CONVENTION AND SHIPPING INDUSTRY REQUIREMENTS

The improvement of the training process is compulsory in the present due to the new position of the

Maritime Education and Training institutions, as providers of services for maritime industry and correspondent activities. In this respect, these institutions have to pay attention to the following underlying factors:

- programmes and courses must meet industry standards and regular requirements;
- programmes and courses must be relevant and meet clients and industry needs;
- training level of graduates must be accordingly with STCW and national authorities requirements;
- teachers and trainers involved in the training process must have a high level of knowledge and understanding of the system and its requirements under present in force regulations.

According with these major objectives, Constanta Maritime University developed its study programmes under requirements of the Convention and applied the curricula recommended by the Convention through the IMO Model Courses for each of the principal specialisations, Navigation and Marine Engine. (Barsan, Hanzu – Pazara, 2007)

Not only the programmes and curricula were developed and updated according with these requirements, also the study cycles were structured in the operational and managerial levels. The training process is designed not only for the students but also for the teachers.

In order to achieve these objectives, Constanta Maritime University started a process of training the trainers, to improve and to update their knowledge and teaching skills to the present conditions and evolutions, based on:

- Development of the lecturers competencies through promotion of knowledge and technologies in the academic maritime field;
- Creation of a development, update and on-line management framework for initial and continuous formative of the human resources;
- Execution of studies and analyses in order to define formative programs and an optimum correlation of these ones with the maritime industry necessities;
- Increase of access and participation of the lecturers to formative programs in order to obtain a double qualification;
- Control of the process and teaching activities through initial and continue formative programs in scope of improvement of TIC using level.

All these are based on the premise that continuous learning is the main condition for reorganization and development of the educational and formative systems, for assurance of decisive competencies during life and to realize the coherency among persons involved in the maritime academic system.

Also, it is necessary to involve maritime lecturers in the international maritime transport framework, to put them in a direct contact with the end users of their work, the companies from the maritime industries and to know exactly their needs. The international maritime companies are the necessary source of information regarding worldwide requests for employment of the maritime personnel.

Collaboration with partners from maritime field, as project objective, is found on communication and information changes in order to identify and implement of adequate modalities to increase the number of work places and to optimize them.

According to the revised STCW Convention, the simulators must be used more effective in the training process of the future seamen and officers. The high technology has to be used in order to increase the level of training and to reach higher standards of knowledge and skills. The use of simulators and technology, especially electronic devices, in the training process offers the possibility to create models close to the reality. As a direct effect, the students are more implicated in the events and also more receptive to the training objectives.

In the first stage, we have to familiarize the students with all of these equipments and make them understand their function and role in the navigation, with implications in the safety of the maritime activities.

Today, many ships are armed with the latest technologies for navigation, as GPS devices, Anti-Collision Radar and Electronic Charts Display, Automatic Identification System devices.

During school training, future operators receive data about technical details, configuration, operational procedures, models of data analysis and correct decisions. During the applications made based on the simulators, the students have the possibility to develop their skills using these devices; they can work with them interconnected, analyze all data or compare data received from two different devices or from other sources. Thus, they will learn to use information in the navigation activities and at the end to realize a safe and correct travel for their virtual ship. (Hanzu-Pazara, Arsenie, Stan, 2008)

Level of skills developed or improved after such training increased in the last years, contributing to an easy access of the Romanian cadets and younger officers to the international maritime work force market. Today, our graduates are accepted as equal competitors with other nationality officers and respected for their knowledge and training level. (Barsan, Hanzu-Pazara, Grosan, 2009)

In the area of the online training, Constanta Maritime University experiences this option inside of a course for familiarization training for petroleum tanker ship operation. Inside of this online course, the students and already certified seafarers interested to attend to a job on a tanker ship, have the possibility to visualize simulated application regarding different operation necessary to be known on a tanker ship, previously, they read and learn the theoretical modules about these.

Analyzing the results, the conclusion is that the students who attend this course have higher knowledge about the operation than the others who did not attend the course. The success is based on the option to see simulated applications and to be familiar with the particular installations and operational procedures characteristically to oil, chemical and gas carrier ships plus the possibility to access from home or from onboard ship the applications during the cadets' practice. (Arsenie, Stan, Surugiu, 2009)

In order to achieve the STCW Convention requirements for onboard training, until 2004, Constanta Maritime University students' practice has been developed onboard of scholarship "Neptun", but due to a lot of engine and hull problems, this activity has been suspended.

After this practice, the solution found by Constanta Maritime University was to send its students in international voyages with different shipping companies, local or international, for this action being contacted the local crewing agencies or owners offices. This was the first step of the current situation, when over half of Constanta Maritime University students covers their requested onboard training on ships of different owners, most of them, international shipping companies with a great rename on the world shipping market, as NYK Ship Management, Japan, Peter Dohle from Germany, Maersk, Denmark, CMA-CGM from France and many others, in totally, 22 shipping companies being part of the partnership. (Barsan, Memet, Stan, 2010), (Hanzu-Pazara, Stan, Grosan, Varsami, 2009)

Taking into account the present regulations regarding onboard training period as cadet, 12 months for deck cadet and 6 months for engine cadets, our University took the decision to help and facilitate students' onboard practice. In this way, in the present agreements and protocols are signed between shipping companies, their local representatives and University, where are stipulated the requested training objectives, onboard live and work condition and schedule for students and the level of theoretical knowledge necessary to be acquired by students before to proceed to the onboard practice.

In time, the number of the shipping companies interested to collaborate with our students increased the number of them increasing, proving the success of the program. During the year of 2008, through this protocol, a number of 555 students covered their onboard practice on ship owned or under management of collaborative shipping companies. (Barsan, Memet, Stan, 2010)

4 CONCLUSIONS

In the present, Constanta Maritime University, as maritime training institution, respects and applies the complete requirements of the STCW Convention and national legislation regarding levels of training and content of the training process according with the final specialisation, deck or engine officer.

Study programmes are structured according with the requirements of the present regulations and with the shipping industry needs, at the end of the study years, the graduates having knowledge and skills necessary to perform their on board duties in respect of the safety and secure procedures and standards.

REFERENCES

Arsenie, P. & Stan, L. & Surugiu, F. 2009. *New development of competencies for younger lecturers according to STCW and training system requirements,* 10th General Assembly of International Association of Maritime Universities – St. Petersburg. Russia. published in MET trends in the XXI century, ISBN 978-5-9509-0046-4, pg. 182-186, Pub. Makarova, Russia.

Barsan, E. & Hanzu-Pazara, R. & Grosan, N. 2009. *The Impact of Technology on Human Resources in Maritime Industry,* 6th International Conference of Management of Technological Changes, Alexandropolis. Greece. Sept. 200. published in Management of Technological Changes, pages: 641-644, ISBN: 978-960-89832-8-1, Publisher: Democritus University of Thrace. Greece

Barsan, E. & Hanzu-Pazara, R. 2007. *New navigation competencies required for an updated STCW Convention,* Journal of Maritime Studies, vol. 21, nr.2/2007, ISSN 13320718. Croatia: 151-161

Barsan, E., Memet, F., Stan,L. 2010. *Particularities of the Maritime Higher Education System as Part of the Maritime Transport Engineering Studies,* 7th WSEAS International Conferance on Engineering Education, Corfu Island, Greece, July 22-24, 2010, published in "Latest Trends on Engineering Education" ISSN: 1792-426X, ISBN: 978-960-474-202-8, Athens, Greece.

Hanzu-Pazara, R. & Arsenie, P. & Stan, L. 2008. *Maritime education and its role in improving safety on the sea,* Proceedings of the 9th Annual General Assembly of The International Association of Maritime Universities „Common Seas, Common Shores: The New Maritime Community", ISBN 978-0-615-25456-4, pp. 165-175, San Francisco. USA. The California Maritime Academy Publisher

Hanzu-Pazara, R. & Stan, L. & Grosan, N. & Varsami, A. 2009. *Particularities of cadets' practice inside of a multinational crew,* 10th General Assembly of International Association of Maritime Universities – St. Petersburg, Russia, ISBN 978-5-9509-0046-4, pg. 99-105, published in MET trends in the XXI century. Pub. Makarova. Russia.

9. Implementation of STCW Convention at the Serbian Military Academy

S. Šoškić, J. Ćurčić & S. Radojević
Serbian Military Academy, Serbia

ABSTRACT: STCW Convention - International Convention on Standards of Training, Certification and Watchkeeping for Seafarers, requires that each member of the International Maritime Organization ensure that education and training objectives and appropriate standards for the competence that should be achieved are clearly defined, and identify levels of knowledge, understanding and training to suit tests and checks required by the Convention. The goals and the appropriate quality standards are clearly defined at the Military Academy of Serbia and can be specified separately for different courses and training programs, measured to meet the administrative requirements of the education system of the IMO, after the training of seafarers.

1 IMPACT OF REPUBLIC OF SERBIA AT THE INTERNATIONAL MARITIME ORGANIZATION (IMO)

International Maritime Organization was developed by the Convention of the Intergovernmental Maritime Consultative Organization (IMCO) at the Maritime Conference of the United Nations in Geneva 1948. In 1982 IMCO changed its name to the International Maritime Organization (IMO)[1]. [1]

The organization now operates as the official organization and specialized agency of the United Nations, based in London. Although established primarily with the task of taking care of the technical aspects of maritime affairs of importance for the safety of maritime navigation, over time, expands its activities in the area of sea protection from pollution, and now takes care of all aspects of harmonization of international maritime activities (among other, maritime security and climate change).

The main body of the organization is the Assembly, which meets every two (usually odd) years. At meetings, resolutions are made which undertake certain organs of the Organization for a certain action, and also resolutions containing recommendations to member states[2].[1] The aim of these resolutions is, first, standardization of state action regarding the safety of maritime navigation and protection of marine pollution. Beside that, members of Assembly elect Council of the Organization, which contains 24 members, tasked to monitor the activities of the organization between the two sessions. Socialist Federal Republic of Yugoslavia was one of the most influential members of the International Maritime Organization (IMO). After the disintegration of SFRY to several new states, membership was only maintained by the Federal Republic of Yugoslavia, later the State Union of Serbia and Montenegro. After the secession of the Montenegro from the State Union of Serbia and Montenegro, membership is retained only by the Republic of Serbia (RS), in which case there was a continuity of membership. Considering that Serbia had her own fleet, from the 80's to the 90's of last century (Belgrade navigation – «Beogradska plovidba-BEOPLOV»)[3], i.e., still has about seven thousand seafarers who sail under the flags of other countries[4], there is a need for Serbia for precise regulation and improvement of relations with the IMO. [2]

[1] IMO has 167 member states today, including the Republic of Serbia. About the organization and actions of the IMO see more at: www.imo.org.

[2] There have been accepted more than 70 maritime conventions and protocols, and over 700 various recommendations and resolutions of the maritime law.

[3] Proceedings of scientific conference "Development and reconstruction of water transport and integration into European transport system, " Traffic Engineering faculty, University of Belgrade: Belgrade, 2002, p. 270.

[4] At Serbian centers for training of seafarers, which work by the STCW Convention of the International Maritime Organization (International Convention on Standards of Training, Certification and Watchkeeping for Seafarers 78/95) and other international treaties, the document in Official gazette of FRY no. 3 on May 11th 2001, about 300 sailors complete the training every year. In our country these jobs are entrusted to Center for training of sailors at the Military Academy of the RS and the School for the shipping, shipbuilding and hydrobuilding.

It's just up to Serbia now, to show and prove that it's able and willing to fully regulate its membership and activities in the IMO in near future, and to fulfill the remaining problems, obligations and requirements in the best possible way[5]. [3]

The public is increasingly mentioning the possibility of establishing a Serbian fleet, following the example of the Swiss merchant navy[6].[4] All this should contribute to promotion and then raising awareness that the desired stronger cooperation of Serbia with the IMO means not only fulfilling regular duties, but the hard work of all mechanisms and actors responsible for compliance with standards and regulations of the International Maritime Organization[7].

2 GUIDELINES OF THE INTERNATIONAL MARITIME ORGANIZATION IN THE IMPLEMENTATION OF INSTRUMENTS

The guidelines have a purpose for members of the IMO to provide resources for the introduction and maintenance measures for successful implementation and enforcement of following IMO Conventions[8]:

- International Convention for the Safety of Life at Sea (SOLAS), 1974, as amended;
- International Convention for the Prevention of Pollution from Ships, 1973, as modified by the Protocol of 1978 relating thereto and by the Protocol of 1997(MARPOL);
- International Convention on Load Lines (LL), 1966; and
- International Convention on Standards of Training, Certification and Watchkeeping for Seafarers (STCW) as amended, including the 1995 and 2010 Manila Amendments.

Under the provisions of the United Nations Convention on the Law of the Sea (UNCLOS) and the above-mentioned IMO Conventions, members are responsible for promulgating laws and regulations and to take other steps as may be necessary to provide full and complete application of these instruments to ensure that, in terms of safety of life at sea and protection of the marine environment, any ship is ready for service for which he was originally designed.

When a convention comes into effect in a country, government must be in a position to enforce its provisions through appropriate national legislative bodies and to provide the necessary infrastructure. This means that government must have a functioning legislative body to pass laws that are applicable to ships under flag of that state, and to ensure their subsequent use, create a possible framework for the implementation of SOLAS, MARPOL, Load Line and STCW Convention by national laws.

In order to successfully meet its obligations, the Republic of Serbia will have:

- to implement the strategy through the promulgating of national laws and guidelines that will assist in the implementation and enforcement of requirements of the Convention;
- to assign responsibilities within administration in charge for updating and revision of strategies as necessary, and
- to formally adopt the above-mentioned in the document for long-term strategic planning.
- Funds which provide fulfillment of requirements of the STCW Convention, as amended. This includes funding to ensure:
- that the training, competency assessment and certification of seafarers are in accordance with the provisions of the Convention;
- that STCW certificates accurately reflect the ability of seafarers serving on any of the ships that sail, using the appropriate STCW terminology as well as terms that are identical to those used in the document on the safe crews, which is issued to a ship;
- that can be carried out impartial investigations of all reported incompetence, act or omissions by the owner of certificates or certificates issued by that Member State, that may pose a direct threat to the safety of life or property at sea or the marine environment;
- that the certificates or verifications issued by members of the IMO may be revoked, suspended or canceled by order or when necessary to avoid abuses;
- that the administrative arrangements, including those relating to training, evaluation and certification are carried out within the scope of another state, such that members accept responsibility for guaranteeing the ability of captains, officers and other seafarers serving on ships. In this regard, particular reference is made to regulation I/2, I/9, I/10 and I/11 of the STCW Convention as amended.

---

[5] The IMO embarked on a policy to encourage developing countries to become members of the IMO and participate in IMO bodies, which were opened to all Member States. Current maritime issues and the International Maritime Organization, Edited by Myron H. Nordquist, John Norton Moore, University of Virginia. Center for Oceans Law and Policy, 1999. Hague, The Netherlands, pp. 391-392.

[6] Switzerland does not have access to the sea, but there is the merchant navy. Switzerland has about thirty ships: ships for bulk cargo, container ships and tankers. The total deadweight tonnage is about one million tons. All ships fly under the flag of the Switzerland.

http://www.swissinfo.ch/eng/news_digest/Merchant_navy.

[7] First of all, the ministry responsible for transport in the Republic of Serbia, i.e., the Ministry of Infrastructure - Department for water traffic, but also part of the Ministry of Foreign Affairs (Directorate for the UN), Ministry of Education (education of seafarers), the Ministry of Health (medical certificate for seafarers), and institutions which are part of centers for training of seafarers.

[8] Refers to the applicable amendments that have been enacted.

National legislation, whether it is a primary or subsidiary legislation, should handle the important issues concerning maritime affairs and one of those questions is practical training, standards of training and certificates of title and certificate. Detailed guidance on this issue is given in the Guidelines for Maritime Legislation, a publication of the United Nations. [5]

3 QUALITY STANDARDS IN TRAINING OF SEAFARERS

In IMO, the role of the human element in safe ship operation and the importance of maintaining high-level training standards for seafarers have long been recognised. The IMO has regularly revised and updated the STCW Convention bearing in mind the importance of the human element in safety management ashore and afloat and in particular, the need to maintain global standard for training for seafarers.[9] [6] By STCW Convention - International Convention on Standards of Training, Certification and Watchkeeping [10], each Member State of the IMO is obligated to provide:

- that all activities related to the training, skills assessment, certification, verification and extension, that are carried out by non-governmental agency or body under their jurisdiction, are under constant supervision of implementation of quality standards to ensure the achievement of goals, including those related to experience of instructors and assessors;
- where government agencies or bodies perform these activities, to implement system of quality standards. [7]

Training for obtaining the authorization for Seafarers (for interested applicants from citizenship) is performed at the Military Academy from June 2007, under International Convention on Standards of Training, Certification and Watchkeeping for seafarers (STCW convention 78/95) and the Rulebook on certificate of competence of the crew members on merchant navy ships.

Group of teachers for navigation and seamanship, conduct training at the Military Academy of Belgrade.

Since the beginning of the organization and execution of training for seafarers, Military Academy is constantly working on quality management system. For the field training courses for seafarers has been established, documented and implemented quality management system, in accordance with the recommendations of ISO 9001:2008 and the STCW Convention. Quality management system was evaluated by an independent certification body **Bureau Veritas**[11].

Our policy of continuous quality improvement of training courses for seafarers at the Military Academy was the permanent commitment of management, leadership and engaged instructors, which is achieved through:

- carrying out training in accordance with the laws of the Republic of Serbia and international regulations and conventions in the field of training of seafarers, and above all harmonized with the Rulebook on certification and competence of the crew members on merchant navy ships, of the FRY and conditions for their acquisition, FRY Official Gazette 67/99 and International Convention Standards of Training, Certification and Watchkeeping for Seafarers 78/95/97/98; performing courses (trainings) and activities with the goal to overcome to meet the needs and expectations of participants in training for acquiring certificates of qualification or special competence of the crew members on merchant navy ships;
- cooperation with the Ministry of Infrastructure (which is in the Republic of Serbia responsible for the implementation of the STCW Convention), port authorities, universities and colleges and other institutions and relevant organizations in the society in the development process of training of seafarers;
- consistent implementation and ongoing improvement of quality management system by teachers and instructors involved, to a complete mastery of quality management and the successful positioning of courses for seafarers training at the Military Academy in the market of education services for seafarers;
- creating a good business relationship and long-term cooperation with business partners and other centers for training of seafarers;
- by permanent monitoring of operating procedures and through process of teaching (training) by using preventive and corrective measures;
- permanent and planning education for all persons involved in the field of seafarers training (attending prestigious training centers for seafarers to receive advanced training according to the STCW Convention 78/95, the presence of teachers and

[9] CONTRIBUTION OF THE INTERNATIONAL MARITIME ORGANIZATION (IMO) TO THE SECRETARY-GENERAL'S REPORT ON OCEANS AND THE LAW OF THE SEA, 2008, MARITIME SAFETY AND SECURITY FUNCTIONS AND CURRENT ACTIVITIES OF IMO AND ITS ACHIEVEMENTS COVERING TECHNICAL FIELDS OF SHIPPING ENGAGED IN INTERNATIONAL TRADE, p.5., http://www.un.org/Depts/los/consultative_process/mar_sec_submissions/imo.pdf

[10] Published in Official gazette - International contracts of FRY May 11th 2001, valid from May 19th 2001.

[11] The Military Academy was issued a certificate № GR13528Q on August 19th 2009.

full-time instructors work procedures on ships of merchant navy);
- establishing and developing business and technical relationships with other institutions of the same or similar activities, thereby contributing to the exchange of experiences and improvement of business;
- entering into new contracts with reputable experts of the field of training of seafarers from the country and abroad, thus expanding the number of experts hired for specific training in the field of maritime affairs;
- seeking to establish the best possible organization of work and plan all key activities and develop a business system,
- reliance of education on the following principles of management: the process approach, focusing on a manual approach in education and training, systematic approach, visionary leadership, decision making based on fact, partnerships with suppliers / vendors, employee involvement, continuous improvement, creating value for students of courses (training), focusing on a value system built in our country, agility in the educational area covered by the Military Academy, autonomy harmonized with the Republic of Serbia.
- efforts to modernize equipment, devices, techniques and procurement of simulators,
- establish a new training under the STCW Convention,
- permanent orientation towards the requirements of candidates for training,
- continuous monitoring of changes and amendments of international conventions related to seafarers' training,
- monitoring and influence on decision making processes of proper and harmonized provisions for training and acquiring certificates of seafarers in the national laws and rulebooks and
- implement a quality as rule rather than coincidence.

Through enterprising policy of quality, based on the previously presented principles and objectives of the course for the training of seafarers, the group of teachers in terms of opportunities and ambient which was created by an environment, tend to make it a place of realizing the interests and serving the needs of candidates, and therefore a place of realizing the interests and needs of marine and river economy. [8]

4 IMPLEMENTATION OF STCW CONVENTION ON MILITARY ACADEMY

The advantage of the Military Academy in courses for training of seafarers in relation to other institutions and organizations in the country and the neighboring area, is possessing its own quality and competent experts in maritime and river traffic, and that Military Academy is equipped with modern equipment, ships, boats, simulators, offices, buildings, swimming pool and polygon, which are necessary and sufficient conditions for the training of seafarers to acquire certificate of competency or a special competence of the crew members of merchant navy.

Entries in the management process of training courses for seafarers are:

- defined quality policy,
- catalogue of courses (trainings) which offers courses (trainings) to students, in accordance with the needs of society, the provisions in relevant laws, decisions of the competent ministries and settings of IMO conventions,
- potential and needs of course participants (training) from the Republic of Serbia and abroad,
- potential and competence at the Military Academy,
- information about the competition that provides the same range of services such as services provided by military academies.

All these entries of the management process collect, sort, check and place the head of courses for seafarers to other competent persons at Military Academy. The head of courses for seafarers cooperate with relevant ministries, all lecturers and instructors of training courses for seafarers, and other organizational units of the Military Academy.

For each course of training of seafarers is prepared a plan for each thematic area (object / subject) at the request of the STCW Convention, rulebook and attachments from international agreements, which states: a brief summary of content processed on the course (subject / topic), area of training, knowledge and skills which are learned during the course (training), wider exposed training program, a method used in training, performance evaluation criteria, teaching aids and devices to be used in training (movies, simulators, training devices, equipment) and literature includes the area for which training is conducted. Thematic plans are updated as changes in the Rulebook or the STCW Convention. [9]

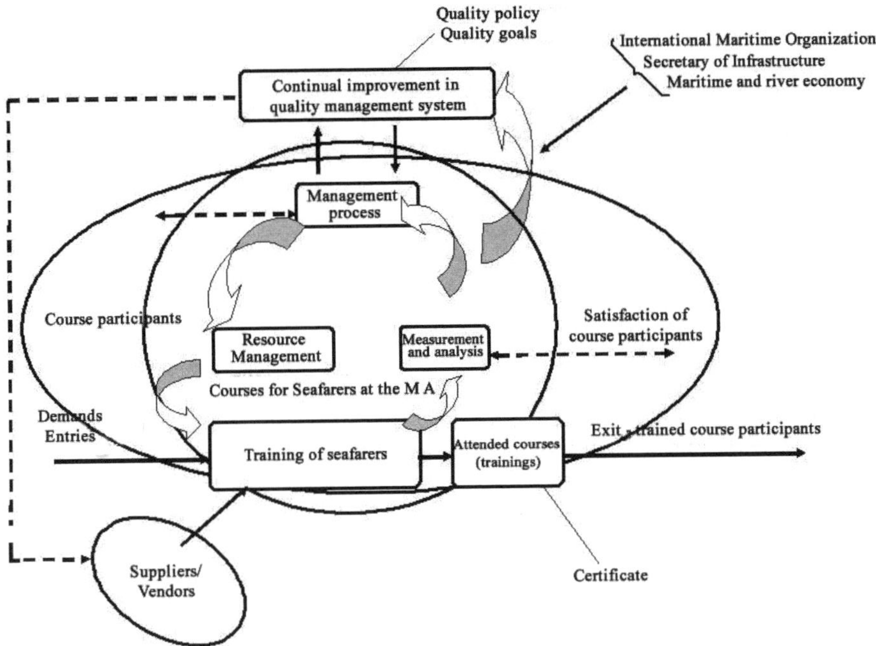

Figure 1. Network of process at seafarer's training course

During the course of training for seafarers at the Military Academy, the tendency is to train all candidates to work on merchant marine ships and to take the exam successfully by providing to the candidates all the necessary knowledge for examinations at the Port Authority.

During the conduct of training, the candidates are allowed immediately and directly questioning the teacher - instructor, in order to clarify the traversed areas and their better preparation, and for all lectures candidates are directed where they can find the literature.

For all areas of training courses for seafarers necessary literature is provided, which is given to all candidates prior to commencement of training.

The Military Academy offers educational materials, devices, simulators, ships, etc.., on which the participants need to perform training courses.

Control of the quality process of making courses is conducted by teachers - instructors, course manager and head of the Military Academy, in accordance with defined procedures, monitoring of course disease, occasionally presence at the process of training, evaluating the quality of performance (course) of training, as required by General specifications for seafarers' training courses and insight into the survey in which the trainees, based on their impressions, evaluate the quality of performance of the course (training) in accordance with defined procedures.

After training, all participants of training, with their signature on the records of attendance at classes and records of training, confirming which topics, intended by training plan, dealt during the presentation of the course (training).

On the basis of decision of the Ministry of defence and decision of the Ministary for capital investments sector for water traffic and navigation safety from February 2007 which approve training for acquariring certificates for training of seafarers, the Military academy provides following certificates:

1 Certificate for familiarization on ship and basic safety training for seafarers (personal survival techniques, fire prevention and fire fighting, elementary first aid, personal safety and social responsibilities); STCW 78/95 A-VI/1 under the following program B1 and B2 of the act standard,
2 Certificate for safety measures on RO-RO ships (Passenger ship safety certificate; RO-RO is not included) STCW 78/95 A-V/3,
3 Certificate for RO-RO passenger ship safety STCW 78/95 A-V/2,
4 Certificate for medical First Aid training STCW 78/95 A-VI/2,
5 Certificate for medical care on board ship STCW 78/95 A-VI/4-1,
6 Certificate for advanced fire fighting STCW 78/95 A-VI/3.

Since June 2007 till the end of 2010, courses for training of seafarers Military Academy has trained in various courses of basic and special training of seafarers, 392 candidates from Serbia and the countries of Southeastern Europe. Center for seafarers training at the Military Academy, insisted on the highest quality of training and acquisition of knowledge and skills of registered candidates. It is the fact that all the candidates successfully passed the exam and that most of them became employees of prestigious shipping companies abroad. The quality of trained personnel for the merchant navy seafarers training center also received several commendations of maritime companies.

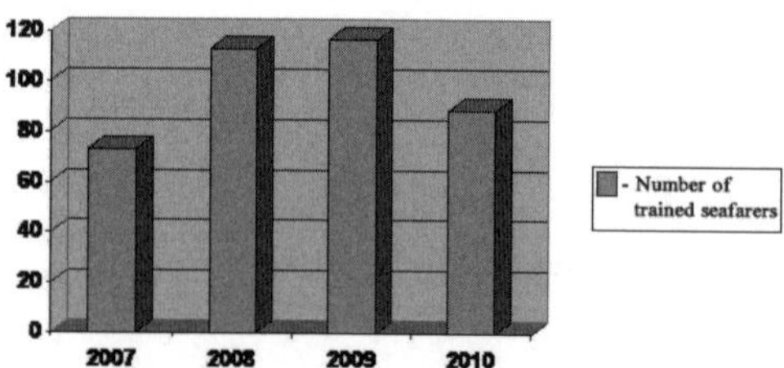

Chart 1. Number of candidates trained by years

5 CONCLUSIONS

The final goals and the appropriate quality standards are clearly defined at the Military Academy of Serbia and can be specified separately for different courses and training programs, measured to meet the administrative requirements of the education system of the IMO, after the training of seafarers. Training for obtaining the authorization for Seafarers (for interested applicants from citizenship) is performed at the Military Academy from June 2007, under International Convention on Standards of Training, Certification and Watchkeeping for seafarers (STCW convention 78/95) and the Rulebook on certificate of competence of the crew members on merchant navy ships.

REFERENCES

[1] http://www.imo.org.,

[2] Development and reconstruction of water transport and integration into European transport system, Transport and Traffic Engineering of the University of Belgrade: Belgrade, 2002,

[3] Current maritime issues and the International Maritime Organization, Myron H. Nordquist, John Norton Moore, University of Virginia. Center for Oceans Law and Policy, 1999. Hague, The Netherlands,

[4] http://www.swissinfo.ch/eng/news_digest/Merchant_navy,

[5] Ljubinka Radosavljevic, Slobodan Milivojevic, Predrag Jevremovic, the International Maritime Organization requirements for inclusion of quality standards necessary to implement the STCW Convention, Belgrade, 2010.,

[6] http://www.un.org/Depts/los/consultative_process/mar_sec_submissions/imo.pdf,

[7] International treaties, the document in Fig. FRY no. 3 / 2001, Belgrade, 2001.,

[8] International Convention on Standards of Training, Certification and Watchkeeping for Seafarers - International Convention on Standards of Training, Certification and Watchkeeping for Seafarers 78/95,

[9] Quality Management System Courses for seafarers' training at the Military Academy, Belgrade, 2009.

10. "Electrical, Electronic and Control Engineering" – New Mandatory Standards of Competence for Engineer Officers, Regarding Provisions of the Manila Amendments to the STCW Code

J. Wyszkowski & J. Mindykowski
Gdynia Maritime University, Gdynia, Poland

ABSTRACT: The paper presents the new requirements for the certification of watchkeeping engineers, chief engineer officers and second engineer officers, in the part related to the function "Electrical, electronic and control engineering", provided in the Manila amendments to the part A of the STCW Code and the consequences for maritime education and training resulting from them.

1 INTRODUCTION

These new requirements for engineers were proposed by Japan, and comparing them to the former text of the STCW'95 Code, now are much more higher.

The reason why Japan decided to do it was presented in its document STW 40/7/54: "*Japan believes that the proposed personnel like an electric officer and an electro-technical officer are not necessary, as long as the existing requirements and qualifications are appropriately maintained. Japan considers that there is a background for the proposals that the requirements in tables A-III/1 and A-III/2 lack detail or specifics and do not seem to reflect the contemporary technologies. Accordingly, Japan proposes amendments to tables A-III/1 and A-III/2, describing the requirements in more concrete ways and adding requirements in K.U.P. (Knowledge, Understanding and Proficiency) regarding high voltage installations that are considered as not a new technology, but a new category (STW 40/7/18)*".

This approach is rather controversial and many opposite opinions were presented and discussed [1], [2]. The IMO STW forum decided to reject Japanese point of view as a general concept, but accepted a development of new "Electrical, electronic and control engineering" standards for engineer officers.

A short comparison of requirements and related standards for electro-technical and engineer officers leads to conclusion, that it would be very difficult or even almost impossible to achieve these new standards for marine engineers.

The authors are of the opinion that above mentioned amendments have a great impact on and must significantly change the model courses for marine engineers.

2 THE FUNCTION "ELECTRICAL, ELECTRONIC AND CONTROL ENGINEERING" IN THE PART A OF THE STCW CODE

2.1 *STCW'95 Code*

Table A-III/1. Specification of minimum standard of competence for officers in charge of an engineering watch in a manned engine-room or designated duty engineers in a periodically unmanned engine-room [3]

Function: electrical, electronic and control engineering at the operational level

Column 1	Column 2	Column 3	Column 4
Competence	Knowledge, understanding and proficiency	Methods for demonstrating competence	Criteria for evaluating competence
Operate alternators, generators and control systems	*Generating plant* Appropriate basic electrical knowledge and skills Preparing, starting, coupling and changing over alternators or gen-	Examination and assessment of evidence obtained from one or more of the following: approved in-service	Operations are planned and carried out in accordance with established rules and procedures to ensure safety of operations

	erators Location of common faults and action to prevent damage *Control systems* Location of common faults and action to prevent damage	experience approved training ship experience approved simulator training, where appropriate approved laboratory equipment training	

Table A-III/2. Specification of minimum standard of competence for chief engineer officers and second engineer officers on ships powered by main propulsion machinery of 3,000 kW propulsion power or more [3]

Function: electrical, electronic and control engineering at the management level

Column 1	Column 2	Column 3	Column 4
Competence	Knowledge, understanding and proficiency	Methods for demonstrating competence	Criteria for evaluating competence
Operate electrical, electronic and control systems	*Theoretical knowledge* Marine electrotechnology, electronics and electrical equipment Fundamentals of automation, instrumentation and control systems *Practical knowledge* Operation, testing and maintenance of electrical and electronic control equipment, including fault diagnostics	Examination and assessment of evidence obtained from one or more of the following: approved in-service experience approved training ship experience approved simulator training, where appropriate approved laboratory equipment training	Operation of equipment and system is in accordance with operating manuals Performance levels are in accordance with technical specifications
Test, detect faults and maintain and restore electrical and electronic control equipment to operating condition		Examination and assessment of evidence obtained from one or more of the following: approved in-service experience approved training ship experience approved simulator training, where appropriate approved laboratory equipment training	Maintenance activities are correctly planned in accordance with technical, legislative, safety and procedural specifications The effect of malfunctions on associated plant and systems is accurately identified, ship's technical drawings are correctly interpreted, measuring and calibrating instruments are correctly used and action taken are justified

2.2 *STCW'2010 Code*

Table A-III/1. Specification of minimum standard of competence for officers in charge of an engineering watch in a manned engine-room or designated duty engineers in a periodically unmanned engine-room [4]

Function: electrical, electronic and control engineering at operational level

Column 1	Column 2	Column 3	Column 4
Competence	Knowledge, understanding and proficiency	Methods for demonstrating competence	Criteria for evaluating competence
Operate electrical, electronic and control systems	Basic configuration and operation principles of the following electrical, electronic and control equipment: .1 electrical equipment: generator and distribution systems preparing, starting, paralleling and	Examination and assessment of evidence obtained from one or more of the following: approved in-service experience approved training ship experience approved simulator training, where	Operations are planned and carried out in accordance with operating manuals, established rules and procedures to ensure safety of operations Electrical, electronic and control systems can be understood and explained with drawings/instructions

	changing over generators electrical motors including starting methodologies high-voltage installations sequential control circuits and associated system devices .2 electronic equipment: characteristics of basic electronic circuit elements flowchart for automatic and control systems functions, characteristics and features of control systems for machinery items, including main propulsion plant operation control and steam boiler automatic controls .3 control systems: various automatic control methodologies and characteristics Proportional–Integral–Derivative (PID) control characteristics and associated system devices for process control	appropriate approved laboratory equipment training	
Maintenance and repair of electrical and electronic equipment	Safety requirements for working on shipboard electrical systems, including the safe isolation of electrical equipment required before personnel are permitted to work on such equipment Maintenance and repair of electrical system equipment, switchboards, electric motors, generator and DC electrical systems and equipment Detection of electric malfunction, location of faults and measures to prevent damage Construction and operation of electrical testing and measuring equipment Function and performance tests of the following equipment and their configuration: .1 monitoring systems .2 automatic control devices .3 protective devices The interpretation of electrical and simple electronic diagrams	Examination and assessment of evidence obtained from one or more of the following: approved workshop skills training approved practical experience and tests approved in-service experience approved training ship experience	Safety measures for working are appropriate Selection and use of hand tools, measuring instruments, and testing equipment are appropriate and interpretation of results is accurate Dismantling, inspecting, repairing and reassembling equipment are in accordance with manuals and good practice Reassembling and performance testing is in accordance with manuals and good practice

Table A-III/2. Specification of minimum standard of competence for chief engineer officers and second engineer officers on ships powered by main propulsion machinery of 3,000 kW propulsion power or more [4]

Function: electrical, electronic and control engineering at management level			
Column 1	Column 2	Column 3	Column 4
Competence	Knowledge, understanding and proficiency	Methods for demonstrating competence	Criteria for evaluating competence
Manage operation of electrical and electronic control equipment	*Theoretical knowledge* Marine electrotechnology, electronics, power electronics, automatic control engineering and safety devices Design features and system configurations of automatic control equipment and safety devices for the following: main engine generator and distribution system steam boiler Design features and system configurations of operational control equipment for electrical motors Design features of high-voltage installations Features of hydraulic and pneumatic control equipment	Examination and assessment of evidence obtained from one or more of the following: approved in-service experience approved training ship experience approved simulator training, where appropriate approved laboratory equipment training	Operation of equipment and system is in accordance with operating manuals Performance levels are in accordance with technical specifications
Manage troubleshooting restoration of electrical and electronic control equipment to operating condition	*Practical knowledge* Troubleshooting of electrical and electronic control equipment Function test of electrical, electronic control equipment and safety devices Troubleshooting of monitoring systems Software version control	Examination and assessment of evidence obtained from one or more of the following: approved in-service experience approved training ship experience approved simulator training, where appropriate approved laboratory equipment training	Maintenance activities are correctly planned in accordance with technical, legislative, safety and procedural specifications Inspection, testing and troubleshooting of equipment are appropriate

2.3 *Analysis of the text of the STCW'95 and STCW'2010 Code*

Comparing specification of minimum standard of competence for officers in charge of an engineering watch listed in Table A-III/1, it is possible to find in the STCW'95 Code only one standard: "Operate alternators, generators and control systems", in the STCW'2010 Code there are two:
"Operate electrical, electronic and control systems" and "Maintenance and repair of electrical and electronic equipment".

In the STCW'95 Code these two standards of competence belonged to chief engineer officers and second engineer officers. It means that in the STCW'2010 Code the requirements for watchkeeping engineers are much higher than before. The contents of the second column (K.U.P.) of the table A-III/1 in the STCW'95 and STCW'2010 Code shows how big is the difference.

The chief engineer officers and second engineer officers in the STCW'2010 Code have two standards of competence:

"Manage operation of electrical and electronic control equipment" and "Manage troubleshooting, restoration of electrical and electronic control equipment to operating condition". It means that now they are not obliged to operate or maintain and repair of electrical, electronic and control equipment, as it was before, now it is a job of watchkeeping engineers.

It is interesting to compare the competences and K.U.P.s of engineer officers and electro-technical officers in the wake of the Manila amendments to the STCW Code [4].

A short comparison of competencies leads to conclusion, that engineer officers are obliged to know how to operate, maintain and repair all electrical, electronic and control systems onboard the ship.

The electro-technical officers should know how to monitor, maintain and repair the systems mentioned above and to operate only generators and distribution systems below and in excess of 1000 V.

With regard to overview of K.U.P.s of both kinds of officers under consideration results that for competence "Operate electrical, electronic and control systems" the requirements listed in K.U.P. column of Table A-III/1 for engineer officers are more detailed than for electro-technical officers.

Concluding, engineer officers competencies and K.U.P.s are comparable and sometimes more exactly described in detailed aspects than appropriate requirements for electro-technical officers.

That's why the validation of existing model training courses:

- 7.02 Chief and 2nd Engineer Officer,
- 7.04 Officer in Charge of an Engineering Watch

is very important and should take into account all these new requirements provided in the STCW'2010 Code.

The authors are of the opinion, that it would be very difficult to meet these requirements without significant increase in the duration of the model courses, especially of the last one.

REFERENCES

[1] Are engineers getting the electrical training they need? In Marine Engineering Review, March 2006, p. 35-36.

[2] Wyszkowski J. et al. 2009. Novelties in the development of the qualification standards for Electro-Technical Officers under STCW Convention requirements. In proc. 8th International Navigational Symposium on Marine Navigation and Safety of Sea Transportation, Trans-Nav, Gdynia, 2009.

[3] STCW Convention. Final Act of the 1995 Conference of Parties to the International Convention on Standards of Training, Certification and Watchkeeping for Seafarers, 1978.

[4] STCW CONF.2-DC-2 - Adoption of the final act and any instruments, resolutions and recommendations resulting from the work of the conference. Draft resolution 2. Adoption of amendment to the seafarers' training, certification and watchkeeping (STCW) Code, 2010.

11. Assessment Components Influencing Effectiveness of Studies: Marine Engineering Students' Opinion

I. Bartusevičienė
Lithuanian Maritime Academy, Klaipeda, Lithuania

L. Rupšienė
Klaipėda University, Klaipeda, Lithuania

ABSTRACT: The phenomenon of effectiveness of studies becomes a hot issue because of its close connection to the quality of studies. Assessment of students' achievements is one of the factors influencing effectiveness of studies; and it is rather problematic nowadays. Theoretical analysis revealed components of students' achievements assessment positively influencing effectiveness of studies if used properly: frequency of assessment events, assessment methods, feedback information provided to students, students' involvement into assessment process (self-evaluation). The questionnaire survey of 132 marine engineering day-time students was performed in 2010. The research results confirmed theoretical insights about the positive influence of properly used students' achievements assessment components (such as assessment frequency, methods of assessment, feedback characteristics, students' involvement into the assessment process) to the effectiveness of studies.

1 INTRODUCTION

Nowadays more and more attention is given to the problem of the quality of higher education. The education quality problem is closely related to the problem of effectiveness of studies, since effectiveness as a quality assessment criterion can be used for assessment and improvement of the performance of educational organizations *(Kokybės vadybos sistemos* 'Quality Management Systems', 2001). At the same time, the quality of organization performance can be increased by enhancing the effectiveness. That is the reason for increased attention of scientists to the problem of educational effectiveness.

The effectiveness of studies is a multidimensional and complex phenomenon. It is influenced by multiple factors; one of them is assessment of students' achievements (Dochy and McDowell, 1997; Black and Williams, 1998; Ramsden, 2003; Butcher, Davies and Highton, 2006; Biggs and Tang, 2007; Sliujsmans, Straetmans and Memenboer, 2008). The phenomenon of the student achievements assessment itself is complicated. In the higher education didactics, it is considered as a rather problematic part of study process. Frequently it does not satisfy teachers and administration of higher educational institution (HEI); it is criticized by students (Black and William, 1998). Therefore, the actual problem exists on a practical level: how the assessment of students' achievements can be organized in order to positively influence the effectiveness of studies, and affect quality of studies.

The assessment problem has been actively investigated worldwide (Marton and Säljö, 1997; Black and Williams, 1998; Ramsden, 2003; Riordan, 2005; Butcher, Davies, Highton, 2006; Biggs and Tang, 2007; Wolf, 2007, etc.). In Lithuania, different aspects of the assessment problem were described by L. Rupšienė and I. Bartusevičienė (2009), etc. Recently, many facts about characteristics of assessment of students' achievements positively influencing the effectiveness of studies are revealed.

Although, it can be pointed out that the students' opinion about the influence of the organization of assessment of their achievements on the effectiveness of studies was not yet under investigation.

The positive influence of assessment to the effectiveness of studies was revealed in several studies (Herman, Osmundson, Ayala, Schneider and Timms, 2006; Nicol and Macfarlane-Dick, 2006; Wolf, 2007; Barr and Tagg, 1995, etc.). For example, S. Yeh's (2008) research results show that the system of quick assessment (when students are assessed from 2 to 5 times per week and immediate feedback is provided) enhances students achievements. Different investigations (Merkhofer, 1954; Anthony, 1967; Stephens, 1951, ref. Bloom, Madaus and Hastings, 1981) proved the positive effect of periodic assessment (in the forms of tests, surveys, etc.) supported by feedback on the results of learning. D. Nicol and D. Macfarlane-Dick (2006); B. Bloom,

G. Madaus and J. Hastings (1981) researches revealed that frequent assessment helps to enhance students' learning motivation and self-esteem. J. Herman, E. Osmundson, C. Ayala, S. Schneider and M. Timms (2006), P. Wolf (2007) proved that the information collected during frequent assessment about students' achievements and learning problems helps teachers to correct teaching and learning process. P. Wolf (2007) highlighted, that frequent assessment allows pointing out teachers' and students' attention to the most important parts of study program; if the learners are used to be assessed they acquire very important additional assessment skill, which will be valuable in future; the assessment helps to develop critical thinking and long life learning strategies. The analysis of ideas of mentioned and other scientists lets ascertain that several components of assessment of students' achievements are especially important: frequency of assessment, assessment methods, feedback characteristics, and self-evaluation.

In the context of shift of teaching/learning paradigm in higher education, the process of assessment of students' achievements and assessment methods themselves are changing. Long time the summative assessment dominated in higher education; which was aimed to measure achievements, evaluate students' performance by grades (marks), classify students using grades, and later issue diplomas and certificates (Butcher, Davies and Highton, 2006). When the traditional summative assessment is used, students' efforts are directed only to pass examination (test). Students are eager to get higher grade (mark) while performing any assessment task. Such assessment does not support students' learning, diminish their learning motivation; even teachers' motivation diminishes if too many assessment tasks are used (Harlen and Deakin Crick, 2003). In the new learning paradigm, much more attention is paid to the learning process, and students' assessment is used for the formative purpose. However, the role of testing in the formative assessment is not reduced, but it is considered as diagnostic tool of students learning problems. The assessment methods having more formative than summative purpose, such as portfolio, laboratory works and researches, problem based tasks and discussions, etc., are used frequently.

The assessment frequency component becomes more apparent in the context of learning paradigm, when the processes of assessment, teaching and learning become integrated, assessment is considered as learning tool (Dochy and McDowell, 1997). Assessment in this case is aimed to develop learning friendly environment and ensure achievement of high students learning results (Barr and Tagg, 1995). In the process of assessment, it is not sufficient to know, that student remembers certain facts at the end of the course, as it was usual in traditional teaching paradigm. The processes of learning and assessment go along; it is considered that students' achievements should be assessed at the beginning, in the learning process and at the end of studying, that is periodically. When arranging frequent assessment of students' achievements, course content is divided into smaller parts, students have to study constantly in order to be prepared for each assessment event. Therefore, at the end of the course students have to study less material for final examination; the tension and stress is avoided, because one failure does not influence academic record of whole course. However, if frequent assessment is not used, students work hardly only before examination, because they postpone learning until the last moment, as a result, the knowledge is acquired only for the short time (Bloom, Madaus and Hastings, 1981). It is stated in the scientific literature (Herman, Osmundson, Ayala, Schneider, Timms, 2006), that it is not required that assessment events has to be performed during every lesson, the intervals between assessment events has not be equal; although, it is desirable to combine all assessment events into the one clear comprehensive system, which is understood by students and teachers.

Students' self-evaluation is defined as feedback provided by the student to him/herself in order to make a decision about his/her performance and effectiveness of studies. Students' self-evaluation is a continuous process (Andrade and Boulay, 2003), helping to enhance learning (Rust, Price and O'Donovan, 2003; Isaksson, 2008; Paris and Paris, 2001), and improve academic results (Andrade and Boulay, 2003; McDonald and Boud, 2003; Irving, Moore and Hamilton, 2003). Self-evaluation may involve different processes, such as self-assessment, self-testing, reflection, etc., aimed to make sound decision.

Feedback can be defined as information provided by a mediator (e.g. academic staff, peer, parent, self, experience) about students' performance and understanding (Hattie and Timperley, 2007). Frequent assessment of students' achievements gives higher results if it is supported by timely positive feedback information (Bloom, Madaus and Hastings, 1981; Black and Williams, 1998; Ramsden, 2003; Nicol and Macfarlane-Dick, 2006, etc.). The investigations revealed problematic aspects of feedback if only grades (marks) are provided. For example, S. Brown and A. Glasner (1999), W. Harlen and R. Deakin Crick (2003) claimed that each test or assessment answer doesn't encourage the process of learning, as students draw all their efforts to getting better marks, thus avoiding the process of real submerging into the subject. This fact has negative influence upon learning motivation. B. Bloom, G. Madaus and J. Hastings (1981) state, that if unsuccessful student's work is always evaluated by low grade (mark), his learning motivation decreases.

However, S. Brown and A. Glasner (1999) noticed that if teachers did not put assessment marks students did not put effort to study more.

It is clear that students' opinion about the influence of the arrangement of the assessment of students' achievements to effectiveness of studies cannot be used to prove hypothesis about impact of assessment to the effectiveness of studies, although it is valuable as additional method besides the other more reliable methods. However, knowledge of students' opinion has theoretical and practical value. From the theoretical point of view, it is important to compare students' opinion to theoretical insights of the assessment issue and ascertain the adequacy of investigated phenomenon, from the practical point of view; students' opinion is valuable for the improvement of assessment of students' achievements aiming to enhance the effectiveness of studies.

Referring to mentioned ideas the problem was formulated for the empirical research: what is students' opinion about influence of assessment of their achievements to the effectiveness of studies? The research was aimed to reveal students opinion about influence of periodic assessment of their achievements to the effectiveness of studies by using the case of marine engineering day-time students.

2 RESEARCH METHODOLOGY

The research population consisted of marine engineering students of Lithuanian Maritime academy in spring 2010. 132 marine engineering day-time students (95 % of all population) in all study years of one mentioned study program were surveyed.

The survey was carried out in May-June 2010. Questionnaires were handed over to students personally. To ensure the ethical question of the research, official permission to carry it out was obtained head of Academy. Besides, all participants of the research were familiarized with the research objective and with special features indicating how to complete the questionnaire. Questionnaires were compiled so that to ensure the participants' anonymity to a maximum; it was impossible to identify the respondent according to the questionnaire's data. Therefore, the essential social research principles were held, namely, voluntarism, anonymity that created such situation when the respondents fully expressed their opinion about the phenomenon or the event of the research.

The originally created questionnaire was used for the research. In the course of theoretical analysis, such components of assessment of students' achievements positively influencing effectiveness of studies if properly used were exposed: 1) frequency of assessment; 2) assessment methods; 3) providing the feedback to the students; 4) students' involvement in the assessment (self-evaluation) (Table 1).

Table 1. The model of students' opinion research.

Generalized students opinion about the influence of assessment to the effectiveness of studies			
Opinion about the frequency of assessment	Opinion about assessment methods	Opinion about providing feedback to the students	Opinion about students' involvement in the assessment (self-evaluation)

The research of the students' opinion about the frequency of assessment. Students were directly asked how many times the teacher has to assess their achievements periodically in order to get the highest effectiveness of studies; they were asked to write a number corresponding to the number of the assessment events during the semester.

The research of the students' opinion about assessment methods. In order to obtain students' opinion the following assessment methods were presented: closed test, open test, answering questions at home, literature review, practical tasks in classroom, practical tasks at home, oral assignment, summary, project preparation, and group discussion. The respondents were asked to choose in the scale of six ranks from "yes for sure" until "not at all" the answer corresponding to their opinion about the influence of every mentioned assessment method to the effectiveness of studies.

The research of the students' opinion about the providing the feedback to the students. The theoretical analysis revealed several assessors able to take part in the assessment process: teacher, student him/herself, and peer students. The question was presented in the questionnaire: which assessor's participation helps to ensure better effectiveness of studies? Several answers were presented for the respondents' choice: 1) teacher; 2) peer student; 3) student him/herself; 4) teacher and student him/herself; 5) teacher and peer student; 6) student him/herself and peer student; 7) teacher, student him/herself and peer student.

The research of the students' opinion about students' involvement in the assessment (self-evaluation). The theoretical analysis revealed that efficiency of the feedback information is determined by several factors: the form of providing information (written, oral), time (immediate, during semester, at the end of course), content (personal improvement, learning problems, knowledge gaps, etc.). The questions were formulated accordingly. In addition, the role of a grade (mark) as a form of feedback information was highlighted in the scientific literature. Therefore, the respondents were asked to provide generalized opinion about influence of grading (marking) of intermediate assignments to the effectiveness of studies. The respondents were asked to mention what grading (marking) strategy better influences the effectiveness of studies: 1) when all presented for assessment assignments are graded (marked); 2) when only written assignments are graded (marked); 3) when only well done assignments are graded (marked); 4) when assignments are not only graded (marked), but also supported by feedback information; 5) when assignments are not graded (marked), but the other feedback information is provided.

The research of the generalized students' opinion about the influence of periodic assessment to the effectiveness of studies. The students were directly asked about the influence of periodic assessment on the effectiveness of studies. In addition, they were asked to mention the model, which in their opinion helps to obtain the highest effectiveness of studies. Three typical assessment models were presented for students' choice: 1) when teacher assesses only during examination at the session time; 2) when teacher gives only one assignment (e.g. summary, course work, control work, etc.) during the semester plus examination; 3) when teacher gives more than one assignment during the semester, which are periodically evaluated, plus examination. Choosing the last answer shows that students recognize influence of periodic assessment to the effectiveness of studies. Moreover, it was aimed to get to know which assessment component is decisive in the context of the effectiveness of studies: frequency of assessment, assessment method, providing feedback information, student's involvement into assessment.

Major qualitative data were analyzed by means of descriptive statistics methods: frequencies, means, standard deviations, minimum and maximum values.

3 RESEARCH RESULTS

3.1 *Students' opinion about the frequency of periodic assessment*

The answers of respondents let make a decision, that the effectiveness of studies would be the highest if students achievements were assessed 3,08 times per semester on the average (minimum value - 1; maximum – 10, standard deviation – 1,660). 32 % of respondents mentioned that the teacher has to assess students achievements 3 times per semester; 21% – 2 times; 17% – 1 time; 14% – 5 times. 1% of respondents pointed that assessment should take part 7 and even 10 times. There was no any respondent having the opinion, that assessment of students' achievements is needless at all.

3.2 *Students' opinion about the assessment methods*

According to the respondents' opinion, all assessment methods are important striving for the highest effectiveness of studies: every method was marked by majority of respondents (from 54% until 85%). The following assessment methods were pointed as effective more frequently: practical tasks in class (pointed by 85 %), oral assignments (73%), project preparation (69%), closed test (67%), summary (67%), group discussion (65%). Several assessment methods as influencing effectiveness of studies were mentioned less frequently, for example, literature review method was mentioned by 53% of respondents, answering questions at home – by 59%, open test – by 62%.

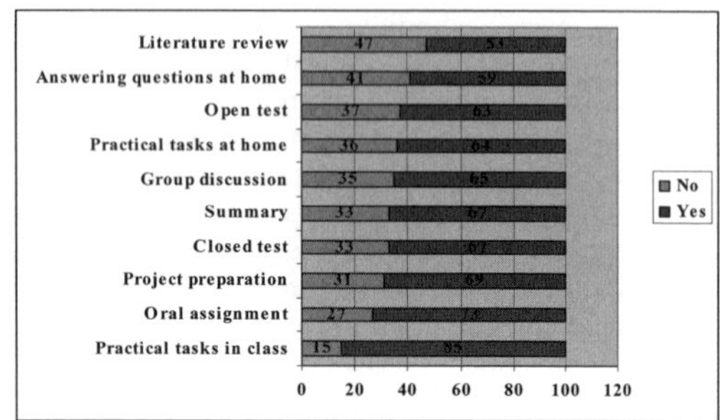

Fig. 2. Students' opinion about influence of assessment methods to the effectiveness of studies, %

The mentioned above data show the importance of chosen by teacher assessment method for effectiveness of studies.

3.3 *Students' opinion about students' involvement in the assessment (self-evaluation)*

The bigger part of respondents (61%) pointed out that the highest effectiveness of studies can be achieved if only teacher assesses students' achievements. One fifth of respondents (21%) mentioned the biggest influence to the effectiveness of studies if following cooperation in the assessment process takes part: teacher and student him/herself. The tendency of acknowledgement of cooperation of all possible assessors (teachers, students and peer students) does not prevail in the answers of respondents: only 6% of respondents pointed out, that participation of teachers, students and/or peers students in the assessment process makes the biggest influence to the effectiveness of studies. Only 3% of respondents think that teacher's and/or peer students' cooperation is the most important (student him/herself does not take part). It is interesting that 4% of students participated in the survey mentioned, that students' achievements could be assessed by student him/herself.

3.4 *Students' opinion about the providing the feedback to the students*

Almost all the students from the research sample (88%) have opinion that assignments have to be marked. 44% pointed out that each interim assessment has to be marked. Some respondents (20%) mentioned that feedback is important along with the mark and 8% of students participated in the survey think, that only providing mark for intermediate assignments is not sufficient for the learning improvement: additional information about students' mistakes, about ways for their work improvement has to be provided by lecturers in oral and written forms.

More that half of respondents (61%) has an opinion that the biggest influence to the effectiveness of studies makes feedback provided in written and oral forms. Only oral feedback information is sufficient for only 25% of respondents, and written information is sufficient for 14% of respondents.

The timing of providing feedback information to students is also important. Half of respondents (50%) mentioned that the biggest influence to the effectiveness of studies is made by providing immediate feedback. More than one third of respondents (35%) has an opinion that feedback information has to be provided during the semester. Only 8% of respondents mentioned, that the feedback has to be provided during examination and 7% did not point importance of feedback for the effectiveness of studies at all.

Analysis of the content of feedback information revealed that the majority of respondents has an opinion, that the effectiveness of studies improves if teachers after intermediate assignment provide following information: tell the students about the important points to learn (75 %), what they have to clarify thoroughly (88%), about personal improvement (80), how the task had to be done (82%), about the gaps in the knowledge and understanding (85%), application of knowledge (88%), how learning had to be performed (78%), about the grade (mark) (75%). Less than half of respondents pointed out, that students have to be compared with each other.

3.5 *Generalized students' opinion about the influence of periodic assessment to the effectiveness of studies*

Majority of respondents (74%) supports the idea, that the effectiveness of studies depends on the assessment model used by teacher. Almost half of respondents (47%) mentioned, that the best effectiveness of studies is achieved if teacher gives more that one assignment during the semester and assesses during examination, 26% of students participated in the survey pointed that study results are the best if only one assignment is done and examination passed, 27% of respondents had the opinion, that the best study results are when teacher assesses only during the examination. Summing up, it can be stated that the majority of students has an opinion that periodic assessment of their achievements positively influences the effectiveness of studies.

Analyzing, which of the assessment components critically influences effectiveness of studies; it was revealed that 37% of respondents give the highest priority to the frequency of periodic assessment events, 23% of respondents – to assessment methods, 23% – to students' involvement into assessment process, and 18% - to feedback.

3.6 *Conclusions*

Majority of students participated in the survey noticed influence of assessment and it's components to the effectiveness of studies. The average of optimal frequency of assessment events was calculated, it was three times per semester. Although some respondents thought, that their achievements could be assessed even more frequently. There were no students mentioned that they do not need assessment at all.

Most students had opinion, that the effectiveness of studies increases if not only traditional assessment methods (such as close and open tests) are used, but new ones, such as group discussions, practical tasks in class and at home, project preparation, answering questions at home, summaries, literature review, oral assignments.

Some of students mentioned advantages of their involvement into assessment process in the context of enhancing the effectiveness of studies. Although this periodic assessment component is not widely used in practice, because majority of all students participated in the survey had an opinion that it is enough when only teacher assesses their achievements.

Almost all students thought, that achieving the highest effectiveness of studies, intermediate assignments have to be graded (marked). One fifth of students mentioned that grade (mark) has to be supported by written and oral feedback information about: the points to address attention; what has to be thoroughly clarified; personal improvement; how the task had to be done; the gaps in the knowledge, understanding, application of knowledge; how learning had to be performed; the grade (mark). The immediate feedback is important for the most students.

Majority of students have an opinion that in general, periodic assessment positively influences the effectiveness of studies. The most important components of periodic assessment are: frequency of assessment events, assessment methods, and providing feedback information.

While analyzing the conclusions of this research in the context of periodic assessment, it can be stated, that Lithuanian undergraduate day-time marine engineering students' opinion is adequate. Analysis results confirm the theoretical insights about the positive influence of students' achievements assessment components, such as assessment frequency, assessment methods, feedback characteristics, and student involvement in the assessment process on the effectiveness of studies if properly used. Therefore, the opinion of students shows, that assessment of students' achievements recently has been transformed from traditional (emphasizing summative assessment) to modern (formative, having motivating character), and the characteristics of the new learning paradigm appears in the assessment process more often.

REFERENCES

Andrade, H. ir Boulay, B. 2003. The role of self-assessment in learning to write. *The Journal of Educational Research, 97(1),* 21-34.

Barr, R. B., Tagg, J. 1995. *From Teaching to Learning - A New Paradigm for Undergraduate Education.* Retrieved from Internet on 16-02-2006: http://critical.tamucc.edu/~blalock/readings/tch2learn.htm#chart.

Bartusevičienė, I., Rupšienė, L. 2009. Periodic assessment of students' achievements as a factor of effectiveness of studies: opinion of social pedagogy students. *Conference Teachers' education in XXI: chantes and perspectives.* Shiauliai: Shiauliai University, 20 November 2009.

Biggs, J.B., Tang, C. 2007. *Teaching for Quality Learning at University.* Berkshire: Mcgraw Hill, Society for Research into Higher Education&Open University Press.

Black, P., Wiliam, D. 1998. Assessment and classroom learning. *Assessment in Education, 5*(1), 7–74

Bloom, B.S., Madaus, G.F., Hastings, J.T. 1981. *Evaluation to Improve Learning.* USA: McGraw-Hill.

Brown, S., Glasner, A. 1999. *Assessment matters in higher education: Choosing and using diverse approaches.* Buckingham: Open University Press.

Butcher, C., Davies, C., Highton, M. 2006. *Designing Learning. From module outline to effective teaching.* London and New York: Routledge Taylor &Francis group.

Dochy, F., Mc Dowell, L. 1997. Assessment as a tool for learning. *Studies in Educational Evaluation, 23* (4), 279-298.

Harlen, W., Deakin Crick, R. 2003. Testing and motivation for learning. *Assessment in Education, 10* (2), 169-207.

Hattie, J., Timperley, H. 2007. The Power of Feedback. *Review of Educational Research;* Mar 2007; 77,1, pp.81-112.

Herman, J.L., Osmundson, E., Ayala, C., Schneider, S., Timms, M. 2006. *The nature and Impact of Teachers' Formative Assessment Practices.* CSE Technical Report. Center for Assessment and Evaluation of Student Learning.

Irving, S., Moore, D., Hamilton, R. 2003. Mentoring for high ability high school students. *Education and Training*, 45, pp. 100-9.

Isaksson, S. 2008. Assess as you go: The effect of continuous assessment on student learning during a short course in archaeology. *Assessment & Evaluation in Higher Education, 33*(1), 1-7.

Kokybės vadybos sistemos. 2001. Pagrindai, terminai ir apibrėžimai (ISO 9000:2000). Vilnius: Lietuvos standartizacijos departamentas.

Marton, F., Säljö, R. 1997. Approaches to learning. In F. Marton, D.Hounsell, & N. Entwistle (Eds.), *The experience of learning.Implications for teaching and studying in higher education,* (second edition, pp. 39-59). Edinburgh: Scottish Academic Press.

McDonald, B., Boud, D. (2003). The impact of self-assessment on achievement: the effects of self-assessment training on performance in external examinations. *Assessment in Education.* 10 (2), 209-220.

Nicol, D.J., Macfariane-Dick, D. 2006. Formative assessment and self-regulated learning: a model and seven principles of good feedback practice. *Studies in Higher Education,31*:2, 199-218.

Paris, S. G., Paris, A. H. 2001. Classroom applications of research on self-regulated learning, *Educational Psychologist,* 36(2), 89–101.

Ramsden, P. 2003. *Learning to Teach in Higher Education.* London: Routledge.

Riordan, T. 2005. *Education for 21st century: teaching, learning and assessment.*

Rupšienė, L., Bartusevičienė, I. 2009. Regulation of Assessment of Learning Outcomes in Lithuanian Higher Schools. *Jaunųjų mokslininkų darbai, 1(22):* 154-162. Shiauliai: Shiauliai University.

Rust, C., Price, M., O'Donovan, B. 2003. Improving students' learning by developing their understanding of assessment criteria and processes. *Assessment and Evaluation in Higher Education.* 28 (2), 147-164.

Sliujsmans, D., Straetmans, G., Memenboer, J. 2008. Integrating authentic assessment with competence-based learning in vocational education: the Protocol Portfolio Scoring. *Journal of Vocational Education and Training, 60(2):*159-172.

Yeh, S. S. 2008. The cost-effectiveness of comprehensive school reform and rapid assessment. *Education Policy Analysis Archives, 16*(13), 1-32.

Wolf, P. J. 2007. Academic improvement through regular assessment. *Peabody Journal of Education, 82*(4), 690-702.

Maritime Education and Training

12. Improving MET Quality: Relationship Between Motives of Choosing Maritime Professions and Students' Approaches to Learning

G. Kalvaitiene, I. Bartusevičienė & V. Senčila
Lithuanian Maritime Academy, Klaipeda, Lithuania

ABSTRACT: Question of improvement quality of studies is a continual hot issue in every educational environment. In maritime education and training, this question is especially important because of international regulations of maritime professions. Quality of studies is a multidimensional and complex phenomenon. It is influenced by wide range of factors: from labour market and current educational policy to individual students' efforts and characteristics. Motives of choosing maritime profession and approaches to learning are characteristics of an individual person. Quality of studies is influenced by students' approaches to learning: deep approach to learning is related with higher quality of studies, surface approach – to lower quality. On the other hand approaches to learning could be influenced by motives of choosing profession, which affect entire professional career planning. Both phenomena are important for the improvement of quality of studies.
The relation between motives of choosing maritime professions and students' approaches to learning in the context of quality of studies are under investigation in this paper.

1 INTRODUCTION

Increasing the quality of maritime studies is a relevant problem, which may be studied in different aspects, and one of them is the study of the students' approach to learning as the individual students' characteristics. Quality of studies is a multidimensional and complex phenomenon (Heywood, 2000; Bartuseviciene, Rupsiene, 2010). The impact of the students' approach to learning on the results of studies was investigated in the works of F. Marton and R. Säljö, (1976), P. Ramsden (2003), N. Petty (2004), G. Pask (1976), N. Entwistle, P. Ramsden (1983). The authors who had investigated the process of studies determined that there were two different students' approaches to learning that were named by the scientists as deep and surface ones. According to those authors different approaches to learning determine different results of learning, therefore, investigating the students' approaches to learning and determining the prerequisites of deep approach to learning that leads to better results of studies, it is possible to find an answer to the question about the increasing of the quality of studies.

J. Biggs (1987a) expanded F. Marton ir R. Säljö (1976) theoretical model by stating that the approach to learning consists of two components: motive of learning and strategy of learning which is understood as a whole of the ways and the habits of learning (Table 1). The construct of students' theoretical approach to learning is based on the idea that the learning motives of students determine the strategies of learning and depend not only on personal characteristics of students but also on learning context and content of learning tasks (Biggs, 1987a). J. Biggs (1987b) created SPQ (*Study Process Questionnaire*) to determine the approach of students to learning, learning motives and strategies, and later on it was revised to a shorter one with 20 questions, R-2F-SPQ (The revised two-factor study process questionnaire*)* (Biggs, Kember, Leung, 2001).

Table 1. Motives and strategies as complex components of the approach to learning (Biggs, 1987b).

Approach	Motive	Strategy
Surface (SA)	Surface motive (SM) is to meet requirements minimally, a balancing act between failing and working more than is necessary	Surface strategy (SS) is to limit target to bare Essentials and reproduce them through learning
Deep (DA)	Deep motive (DM) is intrinsic interest in what is being learned; to develop competence in particular academic subject	Deep strategy (DS) is to discover meaning by reading widely, inter-relating with pervious relevant knowledge, etc.

Analysing the possibilities how to improve MET efficiency of studies it is urgent to investigate the motivation concept of profession choosing and its relation with the approaches to learning that is the prerequisite of high efficiency of studies.

The initiator of the theory of modern choice of professions is considered by F. Parson, who founded the first professional consulting bureau in 1908 in the USA. He formulated the main principles of successful profession choosing (Parson, 1909):

- good self-cognition;
- good knowledge of peculiarities of the chosen profession;
- ability to correctly combine this knowledge and take the right profession solution.

R. Hoppock (1950) explains choosing profession via the satisfaction of need. The essence of his theory is revealed by ten postulates which speak about the fact that a man chooses his profession to satisfy his needs. According to R.Hoppock, choosing profession is being improved when a man starts to imply that the future profession will better satisfy his needs.

J. Holland's theory (1959) is popular among the theoretical and practical people very much, which states that personalities can be divided to six types that were named as realistic, researcher's, artistic, sociable, initiative and normative. In J. Holland's opinion people tend to look for such labour activity environment, where they might express themselves. He states that similar people choose similar professions, but satisfaction from work, success and stability depend on how the personality matches to the environment (Holland, 1966).

The most striking representative of the development model of choosing profession is Donald Super (Super, 1957). The scientist states that by choosing profession a man essentially chooses one of the main means how to express his personal "ego". Professional behavior of a person is a way to implement his professional self-image.

The background of D. Krumboltz theory is learning (1979). According to his statement, there are 4 groups of factors important for the professional self-determination: *genes* – inherited properties, limiting learning possibilities and choosing profession; *environment* – social, cultural, political, economical, natural conditions; *knowledge of learning* – priority development of professions, distribution of certain works, as each individual person has got unique learning knowledge it makes impact on profession choosing; *task fulfillment skills* – there come out task fulfillment standards and values, labour habits, cognition processes and emotional reactions.

According to prof. L. Jovaiša (1999) there are the following motivation factors of profession choosing: social (social state of parents, vicinity of educational institutions), economical (payment for work), psychological (interests, turns, values, intellect and character), health.

The discussed theories should help to understand the factors that determine the solution of profession choosing process. Summarizing scheme of factors, influencing choosing of profession is shown on figure 1.

Tasks of presented investigation are:

- Investigate the motives determining the choice of seafarer's profession.
- Diagnose the individual characteristics of students, their approach to learning that determines the efficiency of studies.

Determine the relations between the motives and approaches to learning.

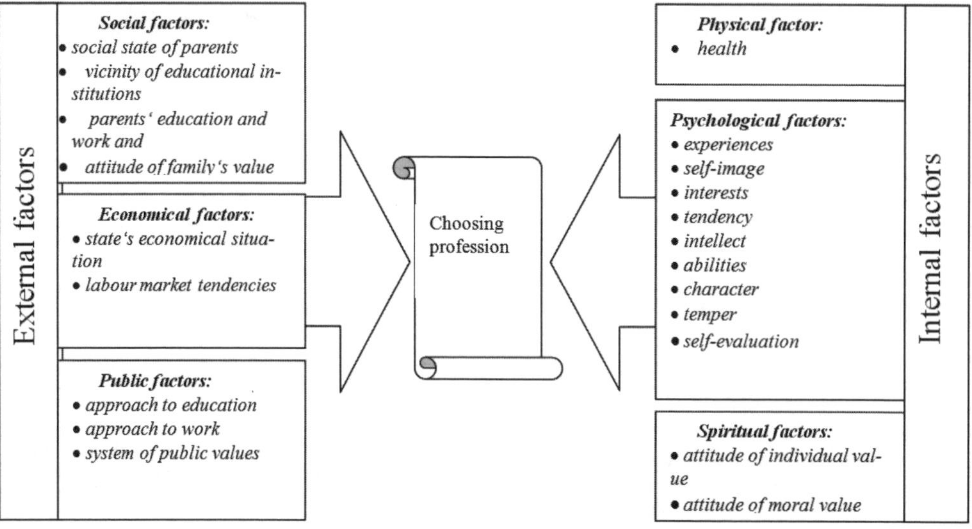

Figure 1. Factors, influencing choosing of profession.

2 THE RESEARCH METHODOLOGY

2.1 *The sample size*

Sample of research was made of full-time students of maritime specialties' studying at Lithuanian Maritime Academy. Making the samples of research the voluntary principle was followed – all the students that were present on query days at school and who expressed their wish were included. Such way of sampling is considered reliable.

In December 2010 – January 2011 233 students from all courses were interrogated (95 % of all maritime specialties' students): first year students – 39,1 percent, second year students – 33,0 percent, third year students – 19,3 percent and fourth year students – 8,6 percent. The sample consisted of 145 students from Marine Navigation study program (62,2 percent) and 88 – Marine Engineering study program students (37,8 percent).

2.2 *The research instrument*

The questionnaire survey was used to collect data in order to examine and verify theoretical and exploratory insights about relationship between motives of choosing maritime professions and students' approaches to learning. The originally developed questionnaire consisted out of 117 questions. The Revised-Two Factor-Study Process Questionnaire (Biggs, Kember, Leung, 2001) translated into the Lithuanian language, adapted, and validated was used as a part of the originally developed questionnaire.

The validity of the R-2F-SPQ questionnaire was checked by confirmatory factor analysis, using VARIMAX method of co-ordinate turning. High KMO ratio (0, 838) and the meaning of Bartlett test (p=0,000) confirmed the suitability of data for factor analysis. During factor analysis four factors were pointed out corresponding subscales of Biggs questionnaire, where the factor weights (L) of components are rather high: from 0,543 to 0,817. Four pointed out factors explained 52,23 percent of variance – such percentage is satisfactory in social sciences (Pett, Lackey, Sullivan, 2003).

2.3 *The data analysis*

The data acquired during the research were analysed using statistical analysis methods (using SPSS for Windows program, 13th version). Analysing quantitative data methods of descriptive statistics (data distribution percents were calculated), non-parametric tests (Mann-Whitney and Kruskal-Wallis tests), factor analysis, correlation tests using ***Spearman's rho*** were applied. For reliability analysis Cronbach's Alpha and Corrected Item-Total Correlation coefficients were used.

3 RESULTS OF INVESTIGATION

3.1 *Analysis of motives of seafarer's profession choosing*

In order to clear out the motives of those young people who chose seafarer's profession the 47 motives were investigated. Having made the analysis it was determined that there were the following important and very much important motives for profession choosing: the seafarer's work is responsible very much (89,7 percent); seafarer's profession is perspective (85 percent); seafarer's profession is masculine (84,2 percent); it is possible to earn well (81,6 percent); seafarers can make career (79 percent); seafarer's work is rather interesting (76,8 percent); seafarers can easily maintain their families (76 percent);74,4 percent wished to get higher education; 74,3 percent of respondents dream to become a captain/navigator or chief mechanic; Seafarers are considered as good specialists (73, 3 percent); seafarer's life is full of adventures (72,9 percent); seafarer's profession is very attractive (70,8 percent); seafarer's profession is one of the most perspective for those who live in seaside region (60,9 percent). 52,6 percent of students who participated in the research considered that when choosing profession an important or very much important motive was the universality of this profession, that after acquisition of seafarer's profession it was possible to work at sea or on shore.

The fact that the respondents purposefully chose seafarer's profession assessing different aspects can be seen in their answers to the statements but only a small part of respondents managed to answer: „I wish to acquire speciality which I am studying but I do not connect my life with the sea" (only 29,2 percent completely or partially agreed to it) and „I wished to study other speciality but I did not manage to enter the institution" (27,9 percent), „I wished to obey my parents" (26,6 percent).

We may come to a conclusion that the motives of choosing profession of the great majority of young people were determined by economical (good salary, possibility to assure social welfare of the family, career possibilities and etc.), social (wish to acquire education, seafarer's work is responsible, seafarers are assessed as good specialists, it is one of the most perspective professions for those who live in seaside region and etc.) and psychological (seafarer's work seems to be interesting, dreams to become a captain or chief mechanic and etc.) factors.

Factor analysis was applied when analysing the students' motivation of choosing seafarer's profession. This method of analysis allows grouping a big quantity of variables, therefore, it was necessary to adapt it for the conditions of this research, as in the questionnaire there were submitted 47 motives of choosing seafarer's profession. When making factor analysis, first of all we had to convince ourselves that the scale of motives is a reliable measuring device and that it is suitable for factor analysis (KMO coefficient 0,852; Barlett's test p meaning - 0,000). Having made factor analysis out of 47 motives of choosing seafarer's profession, 12 factors were picked out that were treated as summarized motives of choosing seafarer's profession (Table 3). The weight of a factor is the correlation coefficient of a factor to the variable. Factor weight meanings are interpreted like all meanings of correlation coefficients. Factors are disclosed by those variables that compose it, the factor weight of which (L) comply with the condition L>0,6. Therefore, when analysing the data, most of the attention is paid to those variables namely.

Table 3. Factor groups of seafarer's profession choosing motives.

Name of a factor	Factor making weights (L_n)
Factor 1. Aptitudes and interests motives	L1
Wished to sail at sea	0,807
Sea always attracted	0,792
From childhood was interested in ships	0,789
Liked everything that was connected with sea	0,747
Wished to try living at sea	0,729
Seafarer's profession was attractive	0,630
Always liked to travel by ships, ferries	0,622
Seafarer's profession seemed the most suitable	0,611
Sea romantics fascinated	0,593
Seafarer's work seems to be interesting	0,584
Was always fascinated by seafarers as strong and brave people	0,542
Factor 2. Career possibilities and economic benefit motive	L2
Think that seafarers can make career	0,789
Seafarer's profession is perspective	0,722
Think that it will be easy to maintain family well	0,635
Seafarer's work is very much responsible	0,611
Factor 3. Accidentally chosen profession motive	L3
Wish to acquire speciality that they learn but do not connect their life with sea	0,783
Seafarer's profession was chosen accidentally	0,777
Wished to study other specialities but did not manage to enter the institutions	0,725
Friends suggested	0,637
Wished to obey parents	0,621
Wanted to avoid military service	0,592
Parents forced	0,553
Factor 4. Influence of relatives motive	L4
Have got a lot of familiar seafarers	0,783
Have got familiar seafarers who might find job in good companies and good salary	0,737
There are seafarers in the family and they wanted to continue family traditions	0,700
Close people advised	0,630
Parents forced	0,533
Factor 5. Emotional attractiveness motive	L5
Lover of nature	0,595
Lithuania needs seafarers	0,590
Liked Seafarer's clothing	0,549
Seafarer's profession is not ordinary	0,516
Factor 6. Benefit from profession motive	L6
Will be able to travel round the world free of charge and see different countries	0,747
Seafarers are honoured in our society	0,532
Factor 7. Masculine profession motive	L7
Seafarer's profession is very much masculine	0,778
Women love seafarers	0,731
Factor 8. Valuable statements motive	L8
Sea teaches to be human being	0,692
Seafarer's profession is one of the most perspective for those who live in seaside region	0,509
Factor 9. Wish to acquire education motive	L9
Wished to get higher education	0,788
Easy to enter maritime specialities	0,617
Factor 10. Professional possibilities motive	L10
Having a speciality that study, they think they will be able to work at sea and on shore	0,770
Think that it will be easy to find a job	0,571
Seafarer's profession is one of the most perspective for those who live in seaside region	0,500
Factor 11. Economic benefit motive	L11
Think that they can earn very well	0,720
Factor 12. Planning to work in maritime business motive	L12
Connect their future with shipping business	0,745

Summarising we may state that maritime profession choosing is determined both by internal and external factors. During factor analysis the following factors are picked out: „Aptitudes and interests motives" and „Emotional attractiveness motive“ correspond to internal factors psychical motives picked out in theoretical model. But factor „Valuable statements motive" and „Masculine profession motive" correspond to internal factors spiritual motives.

Profession choosing was determined by economic motives of external and internal factors. It can be seen by factor groups picked out during the research: „Career possibilities and economic benefit motive“, „Economic benefit motive“, „Professional possibilities motive“, „Planning to work in maritime business motive“. Factor groups: „Wish to acquire education motive “ and „Benefit from profession motive" show that social factors of external factors were very much important in choosing seafarer's profession. Social factors of external factors predetermined choosing seafarer's profession as well. This is confirmed by the picked out factor groups: Influence of relative motive and accidentally chosen profession motive.

3.2 *Relation between the motives of choosing seafarers' professions and approaches to learning*

In order to determine the connection between the motives of choosing seafarer's profession and approaches to learning, it was decided to join each factor defining variables into quantitative variables by aggregation. To assess whether the items which were summed to create aggregated variable, formed a reliable scale, Cronbach's alpha was computed. The internal consistency reliability analysis indicating the consistency of a multiple item scale appropriate for summation of variables to the aggregated variable can be done by using Cronbach's alpha coefficient (Leech, Barrett, Morgan, 2008). It was discovered, that scales of six motives can be used for summation, because Cronbach's Alpha of items of that motives were bigger or close to 0,7 (Table 4).

Table 4. Aggregated variables – factors of motives.

Number	Aggregated variables – factors of motives	Cronbach's Alpha
1	Aptitudes and interests motives	0,914
2	Career possibilities and economic benefit motive	0,779
3	Accidentally chosen profession motive	0,840
4	Influence of relatives motive	0,753
6	Benefit from profession motive	0,696
5	Emotional attractiveness motive	0,677
7	Masculine profession motive	0,551
8	Valuable statements motive	0,410
9	Wish to acquire education motive	0,405
10	Professional possibilities motive	0,452

Six new agregated variables, named according to motive‘s names, were used for correlation analysis of motives of choosing profession and approaches to learning. For correlation analysis Sreaman's coefficient was used, because scores of variables were not normally distributed (Morgan, Leech, Gloeckner, Barrett, 2007).

The correlation values of Spearman's rho presented in the table 5 show statistically significant correlations of Deep Approach scores and emotional attractiveness, aptitudes and interests, benefits from profession motives. This can be understood, that students study more effectively if they have chosen the profession based on their emotions and interests, and understand advantages of chosen profession.

Table 5. Spearman's rho coefficients of statistically significant correlations of Deep Approach to studies and motives.

Number	Aggregated variables – factors of motives	Spearman's rho	p
5	Emotional attractiveness motives	0,316	0,000
1	Aptitudes and interests motives	0,313	0,000
6	Benefits from profession motives	0,186	0,005

The analysis showed statistically significant correlations (p=0,000) of Surface Approach with motives of choosing profession in two cases:

1 with Accidentally chosen profession motive (Spearman's rho=0,432);

2 with Influence of relatives motive (Spearman's rho=0,282).

The investigations can lead to the conclusion that if student chooses the profession accidentally or the decision is influenced by relatives, student uses Surface Approach to learning and his study effectiveness in this case is not very high.

Summing up the results of approaches to learning and motives of choosing professions correlation analysis, it can be stated, that understanding of mentioned relations can directly influence quality of maritime education and training: students' emotions, interests, and understanding of the advantages of maritime professions lead to higher quality of studies, but accidental or influenced by relatives decision to become a seafarer lead to less qualitative studies. It was found, that first year students more than upper course students and female students more than male students are oriented towards Deep Approach to learning.

Statistically proved conclusion can be done that maritime education and training institutions have to explain young people all merits of maritime profession and show possibility for them to find emotional attractiveness and realization of their interests if they will choose maritime professions. In this case, according to the results of the research the studies will be effective and quality of MET will increase.

4 CONCLUSIONS

The motives of choosing profession of the great majority of young people were determined by economical (good salary, possibility to maintain family welfare, career possibilities and etc.), social (wish to acquire education, seafarer's work is responsible, seafarers are valuated as specialists, seafarer's profession is one of the most perspective for those who live in seaside region and etc.) and psychological (seafarer's work seemed to be very interesting, dream to become a captain or chief mechanic and etc.) factors. Both external and internal factors predetermine the choice of seafarer's profession.

The results of correlation analysis of motives of choosing maritime professions' and approaches to learning showed relations of Deep Approach with students' emotions (Spearman's rho=0,316, p=0,000) interests (Spearman's rho=0,313, p=0,000) and understanding of the advantages of maritime professions (Spearman's rho=0,186, p=0,000) that means if student chooses profession following mentioned motives, his studies are of higher quality. Relations of Surface Approach with accidental (Spearman's rho=0,432, p=0,000) or influenced by relatives (Spearman's rho=0,282, p=0,000) motives to take decision to become a seafarer show less qualitative studies.

It is clearly proved statistically that if student chooses profession following his emotions, interest and understanding of the advantages of professions, his studies are more effective, than if he chooses profession accidentally or influenced by relatives. It was found, that first year students more than upper course students and female students more than male students are oriented towards Deep Approach to learning. The conclusion can be done that maritime education and training institutions have to explain young people all merits of maritime profession and show possibility for them to find emotional attractiveness and realization of their interests if they choose maritime professions. In this case, according to the results of the research the studies will be effective and quality of MET will increase.

REFERENCES

Bartusevičienė, I., Rupšienė, L. 2010. Periodic assessment of students' achievements as a factor of effectiveness of studies: the opinion of social pedagogy students. *Tiltai, 2 (51)*, pp. 99-112.

Biggs, J. 1987a. *Student approaches to learning and studying*. Melbourne: Australian Council for Educational Research.

Biggs, J. 1987b. *The Study Process Questionnaire (SPQ)*. Manual. Hawthorn, Vic.: Australian Council for Educational Research.

Biggs, J., Kember, D., Leung, D. 2001. The revised two-factor study process questionnaire: R-SPQ-2F. *British Journal of Educational Psychology*, Mar 2001, 71, pp. 133-149.

Crick, R. D. 2007. Learning how to learn: The dynamic assessment of learning power. *Curriculum Journal 18(2)*: 135–153.

Heywood, J. 2000. *Assessment in Higher education. Student Learning, Teaching, Programmes and Institutions*. London: Jessica Kingsley Publishers Ltd.

Holland, J. L. (1966). The psychology of vocational choice. Waltham, MA: Blaisdell.

Hoppock, R. 1950. *Presidential Address 1950*. Occupations, 28, 497-499.

Entwistle, N., Ramsden, P. 1983. *Understanding Student Learning*. London: Groom Helm.

Krumboltz, J. D. 1979. A *social learning theory of career decision making*. In A. M. Mitchell, G. B. Jones, & J.

Leech, N. L., Barrett, K. C., Morgan, G. A. 2008. *SPSS for Intermediate Statistics: Use and Interpretation*. New York: Taylor & Francis Group, LLC.

Marton, F., Säljö, R. 1997. Approaches to learning. In: F. Marton, D. Hounsell, N. Entwistle (eds.). *The experience of learning. Implications for teaching and studying in higher education*. Second edition. Edinburgh: Scottish Academic Press, 39–59.

Morgan, G. A., Leech, N. L., Gloeckner, G. W., Barrett, K. C. 2007. *SPSS for Introductory Statistics: Use and Interpretation*. New Jersey: Lawrence Erlbaum Associated, Publishers.

Pask, G. 1976. Styles and strategies of learning. *British Journal of Educational Psychology 46*: 128–148.

Parsons, F. 1909. *Choosing a vocation*. Boston: Houghton Mifflin.

Petty, G. 2004. *Teaching Today*. Cheltenham: Nelson Thornes Ltd.

Ramsden, P. 2003. *Learning to Teach in Higher Education*. London: Routledge.

Super, D. E. 1957. *The psychology of careers*. New York: Harper & Row.

13. Evaluation of Educational Software for Marine Training with the Aid of Neuroscience Methods and Tools

D. Papachristos & N. Nikitakos
Dept. Shipping, Trade and Transport, University of Aegean, Greece

ABSTRACT: The evaluation with the use of neuroscience methods and tools of a student's satisfaction – happiness from using the e-learning system (e-learning platforms, e-games, simulators) poses an important research subject matter. In the present project it is presented a research on course conducted in the Faculty of Merchants Officers, Marine Academy of Aspropyrgos. In particular, this research with the use of a neuroscience tool-gaze trucker, investigates the amount of satisfaction of the students using a simulator by monitoring the users' eye movement in combination with the use of qualitative and quantitative methods.

1 INTRODUCTION

The effectiveness evaluation of the Information & Communication Technology (ICT) in the didactics was mainly based on the experience & analysis (positivistic) methods, which accept that knowledge may be attributed only to the objective reality existing regardless of the values and beliefs the ones seeking to discover her. Methods of the physics and behavioral sciences are adopted, as well as objective forms of knowledge and deterministic acknowledgements for the human nature. As it shown in the international bibliography, the use of multiple methods of evaluation is more effective and the combinatorial use of quantitative and qualitative approaches confines their weaknesses (Brannen, 1995, Bryman, 1995, Patton, 1990, Retalis et al., 2005). That combination may bring out multiple applications of the ICT in the educational sector, thus contributing in a more "sufficient" evaluation of an application. According to Hubermas (1971), the final target of the ICT effectivity evaluation on the educational procedure must call for the examination and the evaluation of the three interests on knowledge: the technical (suggests the scientific opinion – experience & analysis example), the practical (the interpretation methods that offer knowledge that serves the "practical" interest on the understanding and interest) and "manumission" (offers the necessary critical and dialectical basis for the substantial connection between theory and action). Knowledge is the nucleus and around should orbit the evaluation procedure as well as the result of the human action that is defined by the physical needs and interests that lead and at the same time shape the way the knowledge is structured in several human activities. It is more than obvious that all the above conclude to the fact that the evaluation targets and the interest for knowledge define the evaluation mode and its results (Carr and Kemmis, 1997, Hubermas, 1971, Retalis et al., 2005).

In the evaluation of the tutorial systems, the system's term "utility" is analyzed in two supplementary components: the utility offered to the final regarding the system's efficiency and the usability regarding the facility of the users to comprehend or use that usability. These two senses are bound together but it's not necessary one to exist without the other (Grudin, 1992). In the evaluation, in particular, is widely used the term "manageability" which is a self-evident requirement for all the systems and tools managed by men (Avouris, 2000, Avouris et al., 2001, Nielsen, 1993, Papachristos et al., 2010).

One of the problems emerging by the evaluation techniques is the fact that they are based on the observation and on the users' answers. Today a great interest is imposed on the application of the objective usability testing. This testing mainly concerns the observation of the eyes and the measurement of physical data related to several part of the human physiology (heart activities, activity of perspiration glens, electric activity of the muscles and brain). Recording the measurements requires the application of several organs and sensors on the users. The existence and correspondence of physiological measurements patterns in specific emotional situations and in general the determination of a theoretical frame of interpretation that defines a user's emotional reaction in an interface is a state of art research field (Dix et al., 2004, Picard, 1997, Retalis et al., 2005).

Today the use of the consolidated strategic research that is called cognitive neuroscience and includes the study of the behavior and the external situations related to it, as well as the expansion of the nervous system mechanisms that intervene in this relationship, leads to a better expansion of the user's physiological reactions. In the marine education and training, in particular, the modelization of the student's satisfaction based on objective standards (biometrical measurements) that present a better frame of interpretation is an important research subject. Via cognitive neuroscience is possible to determine a background that explains the satisfaction effect and at the same time presents new considerations that will expand the existing so far educational conclusions regarding the adults' education. The coexistence of the obtainment and understanding of knowledge and the acquisition of skills (according to specific standards) necessary to practice the marine profession is a crucial element for the marine professionals. In addition it will promote the design improvement of the e-learning software programs and will help the e-learning software developers/manufacturers to improve their products on the terms of the learning success and effectiveness (Goswami, 2007, IMO, 2003, Papachristos and Nikitakos, 2010, Kluj, 2002).

2 RESEARCH METHODOLOGY

The recommended research procedure aims at the modelization of the (subjected) satisfaction of the students – users in marine education via the user interface evaluation of the MATLAB simulator. The experiment is conducted in the Faculty of Merchants Officers, Marine Academy of Aspropyrgos with a random sampling among the students. The experiment's researching purposes concern the evaluation of the students' satisfaction from the MATLAB simulator use and the educational evaluation of MATLAB from the user's point of view.

The research that will be conducted is a combination of a qualitative – quantitative methodology on one hand and a use of neuroscience tools (gaze trucker use) on the other hand. The purpose of that is to combine the positive elements of the corresponding methodologies: targeting at measurable results & variable testing (quantitative), interpretative, explanatory (qualitative) and more objective measurements by "observation" of physiological data of the user (neuroscience tools).

For the quantitative research, the questionnaire method will be applied in order to extract measurable results regarding the educational evaluation of the MATLAB by the users in combination to the elements that satisfy the user by the simulator's operation. The use of the qualitative research via semi-structured interviews was chosen in order to examine the deeper reasons of the users' satisfaction of the simulator and verify the neuroscience tool measurements and the quantitative research results. The use of the gaze trucker was chosen because it presents more "accurate" measurements. The course of the researching procedure in the research conduction is shown in the following figure (Fig.1).

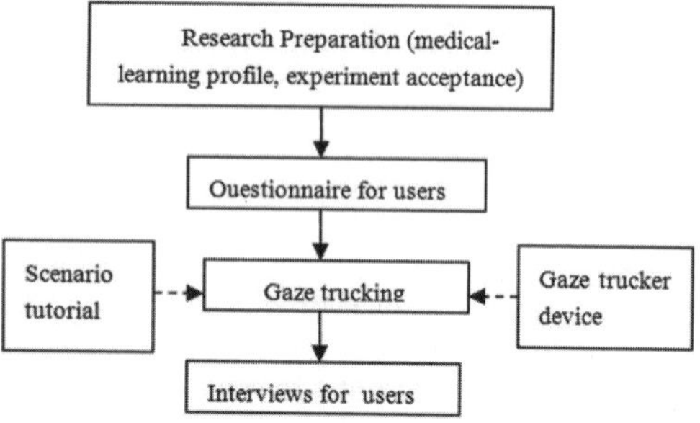

Figure 1. Research steps.

The determination of the satisfaction level concerns the following parameters: system's usability regarding the whole system (total usability), as well as the corresponding learning scenario, the stimulant for the active participation of the users (from the system and the learning scenario) and the user's friendliness regarding interface. The emerging of conclusions from the research results concludes the evaluation procedure that concerns the "identification" of the measurements from the gaze trucker in relation with firstly the questionnaire results (educational evaluation) and secondly and most importantly the interview results (educational evaluation & determination of the user's satisfaction data) (Fig. 2). The evaluation's intention is to create measurement patterns that correspond to psychological characteristics. The learning scenario applied was developed according to the STCW' 95 corresponding educational specifications on the Merchant Officers education in collaboration with the staff of the Marine Academy of Aspropyrgos.

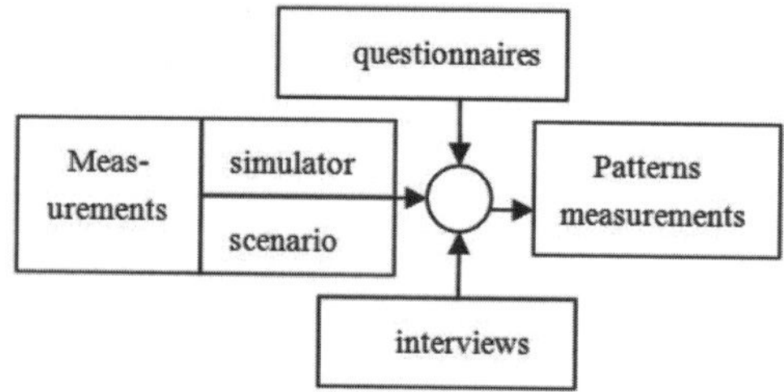

Figure 2. Verification procedure.

3 OPTICAL MEASUREMENT TECHNIQUE

The human vision/eyesight is an extremely complicated activity with physical restrictions as well as restrictions concerning perception. The optical perception is distinguished in the physical reception of a stimulant from the outer world and the processing/interpretation of the certain stimulant. The observation of the gaze trucking as well as the pupil is a possibility existing in many years now but the technological developments in both material equipment and software made it more viable, mostly as an approach to measure usability (Duchowski, 2003, Dix et al., 2004). The eye movements are supposed to illustrate the amount of cognitive elaboration a screen demands and therefore the difficulties and facilities in its processing. In general the optical measurement focuses on the following (Goldberg and Kotval, 1999, Dix et al., 2004): the focus points of the eyes, the eye movement patterns and the variations of the eye pupil. The methodology of effectuation of the optical measurement that was chosen on the present experiment aims at the eye movement observation and the user's head movement regarding with time. For carrying out of the experiment the "Face Analysis" software developed by the Image, Video and Multimedia Systems Lab (IVML) του National Technical University of Athens (NTUA) is applied in collaboration with a Web camera connected to the computer hosting the research subject (MATLAB) (Asteriadis et al., 2009). That software records (parameters): (a) Eye gaze vector: vertical and horizontal movements (2 floats), (b) Head Pose Vector: pitch, yaw (2 floats), pitch and yaw come in normalized floats, (c) Dist_monitor: Float indicating distance of the user from the monitor (~1: fixed distance, <1 goes far, >1 comes close)(1 float) and (d) Head roll: roll comes in degrees (1 float). The whole data recording and analysis procedure is shown in the following figure (Fig.3).

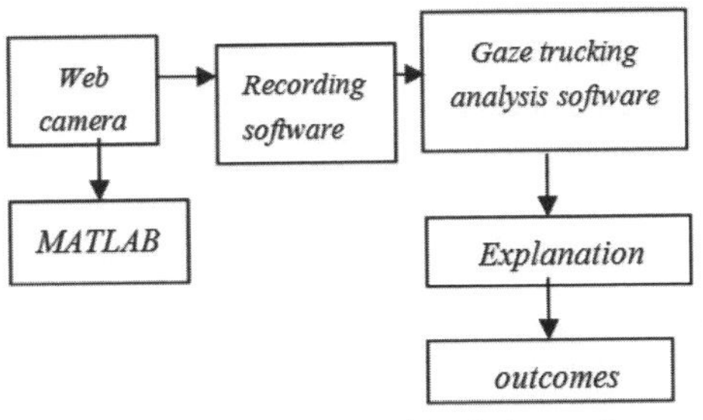

Figure 3. Optical Data Recording and Analysis procedure.

4 FIRST RESULTS

The first (random) sampling was carried out on the January 2011, in the Information Technologies Laboratory of the Marine Academy of Aspropyrgos. The samples consisted of 11 students (9 Male, 2 Female) that were subjected to a specific experimental procedure (research methodology) (Table 1). The sample's learning – medical profile as shown in Table 2 presents homogeneity, whereas from the first processing of the Matlab gaze trucker data emerge useful information (Table 3, 4).

Table 1. Structure of Sample

Variables	Male (2)*	*Female (2)*
Semester	E (9)	E (2)
Pass "Control System & Matlab" course	-	-
Study "Control System & Matlab" course	(9)	(2)
No pass "Control System & Matlab" course	-	-
Level computer using (No Use, Basic, Medium, Advance)	M(6),B(2)A(1)	B(1), M(1)

*(frequency)

Table 2. Structure of medical-learning profile

Question theme	Male (2)*	*Female (2)*
Strabismus	N*(9)	N(2)
Monochromatism	N (9)	N (2)
Eye disease	N (8), Y(1-myopia,astigmatism)	N, Y*(1-myopia)
Eye operation	N(9)	N(2)
Dyslexia	N(8), Y(1)	N(2)
ADHD	N(9)	N(2)

*N: No, *Y: Yes, *(frequency)

Table 3. 1st Data Set of Sample.

	Time (min) (aprox.)	Eye gaze vector vertical	Eye gaze vector horizontal
Average	F*(13.5), M*(8.5), T*(9.4)	F*(0), M*(0), T*(0)	F*(-.16), M*(7.04), T*(6.16)
Standard Dev	F*(2.12), M*(3.3), T*(3.67)	F*(0), M*(0), T*(0)	F*(0.54), M*(31.81), T*(29.7)

F*:Female, M*:Male, T*:Total

Table 4. 2nd Data Set of Sample.

	Dist_monitor	Head roll	Head pose vector pitch	Head pose vector yaw
Average	F*(1.2), M*(1.23), T*(1.22)	F*(-1.2), M*(2.03), T*(1.14)	F*(0.04), M*(0.04), T*(0.04)	F*(-0.16), M*(-0.04), T*(-0.07)
Standard Dev	F*(0.29), M*(1.07), T*(0.92)	F*(7.90), M*(10.12), T*(9.67)	F*(0.31), M*(0.25), T*(0.21)	F*(0.54), M*(0.35), T*(0.41)

F*:Female, M*:Male, T*:Total

The first measurements show that men needed less time to execute the scenario than women (the interview showed that women are more informed about Matlab and the control systems and answered in more questions than men). In the distance from screen observation (dist_mon variable) it was recorded that both men and women approach the screen (>1), which means that they both fully watching the scenario and have difficulty in using Matlab. As the screen distance increased in men, they seemed to have a more attention of the scenario and the Matlab use, even though they don't seem to achieve the same success in the scenario solving case as the women. In the next variable, we observe a non-balance in the inclination of the head (~0) in both sexes, which probably suggests a deeper inspection of the Matlab interface (that is probably ought to the lack of more practice time). The eye gaze vector approximately shows (the females marginally watch inside the screen although the sample is small), the students watch inside the screen and not outside of it (there is an interest on the scenario execution with the help of Matlab software or search about that).

The interview shows that 7 people from a total of 11 consider that Matlab software improves teaching the control systems (6 male, 1 female), all of them consider the Matlab interface to be friendly to the user and that it is necessary to have a basic mathematics background in order to use it. Without a mathematics background, its operation is difficult. During the execution of the learning scenario (closed control system solving exercise and calculation of time response with a paced entrance) none of the 11 students could get to a point of solution, rendering it to the lack of necessary time practice of Matlab in the School's educational program and 4 students (4 male) to the fact that they are not interested in the control systems & Matlab material. Furthermore 2 people from 11 (1 man, 1 woman) thought that the scenario solution was difficult due to the difficulty in Matlab use. In total 7 people from 11 agreed that the test (scenario) conducted in Matlab in the control systems course is satisfying, whereas 4 of them thought otherwise (3 male, 1 female).

5 DISCUSSION

The scientific approach regarding the study of psychological phenomena has wider application in the fields of education, since a basic objective of the educational process is to facilitate the process of learning and of Neuroscience, is the study of the nervous system as a mediator of behaviour. The evaluation using neuroscience methods and tools of happiness, satisfaction of the trainee after the use of electronic learning systems (e-learning platforms, e-games, simulators) is an important issue of research. Furthermore in the marine education, the use of more objective evaluation methods of the modern educational environments will offer a qualitative improvement of the educational programs and at the same time the re-development of the e-learning methods that will benefit both the students and the educational process (Blakemore and Frith, 2005, Goswami, 2007, Pare – Blagoev, 2007).

Qualitative upgrading of the educational process at university level marine education depends largely on the instructive value of the trainers' educational software. Marine education software has developed out of a specific initial implementation in an equivalent manner to which the programs of practice and training have been applied in cases of simulations and programming environments. In most cases, educational software categories have developed without taking into consideration any special pedagogical theory (Tsoukalas et al., 2008).

In the present study, the evaluation of educational software (simulator type) is conducted, in particular MATLAB, which is used in the laboratory of control systems of the engineers of the Merchant Navy from the learners' perspective. The research takes place in the School of Engineering of the Merchant Marine Academy of Aspropyrgos in a first sample of 11 students (target:21 students). Specifically, a biometric tool of visual recording (a neuroscience tool) records the trainees' satisfaction from the educational use of the simulator, by watching the users' eye movements in conjunction with questionnaires & interviews (Photo 1, 2).

6 CONCLUSIONS

A research that offers an objective data recording concerning the emotional state of a user that affects his/her problem solving ability and carrying out of projects poses an important challenge. Overall the first results indicate the following so far: (a) possible relationship between the scenario comprehension and the Matlab use with the head being in a distance from the screen, (b) the head's inclination shows the expansion of the Matlab use among the users, indicating an increase in training time of the Matlab use (modification of the educational program and the corresponding standard STCW'96) and (c) possible relationship between the eye gaze vector (screen monitoring) and the users' satisfaction (interest = satisfaction).

The research continues with the numeral increase of the sample and the total processing and evaluation of the research findings (qualitative and quantitative data).

REFERENCES

Asteriadis, S. Tzouveli, P. Karpouzis, K. Kollias, S. 2009. Estimation of behavioral user state based on eye gaze and head pose—application in an e-learning environment, *Multimedia Tools and Applications*, Springer, Volume 41, Number 3 / February, pp. 469-493.

Avouris, N. Tselios, N. Tatakis, E. C. 2001. Development and Evaluation of a Computer-based Laboratory teaching tool. *Journal Computer Applications in Engineering Education*, 9(1), 8-19.

Avouris, N. 2000. *Human-Computer Interaction:An Introduction*. Athens:Diavlos (in greek).

Blakemore, S. J. Frith. U. 2005. *The Learning Brain: Lessons for Educations*. Oxford:Blackwell.

Brannen, J. 1995. Combining qualitative and quantitative approaches: An overview, J. Brannen (ed.), *Mixing Methods: Qualitative and Quantitative Research*. UK:Avebury, 3-38.

Bryman, J. 1995. Quantitative and qualitative research:further reflections on their integration, *Mixing Methods: Qualitative and Quantitative Research*. UK:Avebury, 57-80.

Dix, A. Finlay, J. Abowd, G. D. Beale, R. 2004. *Human-Computer Interaction*, UK:Pearson Education Limited.

Duchowski, A. T. 2003. *Eye Tracking Methodology:Theory and Practice*, London: Springer-Verlag.

Habermas, J. 1971. *Knowledge and Human Interest*. USA:Beacon Press.

Goldberg, J. H. & Kotval, X. P. 1999. Computer interface evaluation using eye movements:methods and constructs. *International Journal of Industrial Economics*, 24:631-45.

Goswami. U. 2007. Neuroscience and education:from research to practice? *Nature Review Neuroscience*, 7:406-413.

Grudin, J. 1992. Utility and usability: Research issues and development contexts. *Interacting with Computers*, 4(2), 209-217.

IMO-International, Maritime Organization, 2003. Issues for training seafarers resulting from the implementation onboard technology, *STW 34/INF.6*.

Kluj, S. 2002. Relationship between learning goals and proper simulator, *ICERRS5 Paper*.

Nielse, J. 1993. *Usability Engineering*. UK:Academic Press.

Papachristos, D., Alafodimos, N., Zafeiri, E., Sigalas, I. 2010. Educational evaluation of Matlab simulation environment in teaching technological courses:The example of Digital Control Systems, *Proceedings of the Hsci2010 - 7th International Conference on Hands-on Science,* The University of Crete campus at Rethymno, July 25-31, 2010, pp.168-174, http://www.clab.edc.uoc.gr/HSci2010/.

Papachristos, D. Nikitakos, N. 2010. Application Methods and Tools of Neuroscience, in Marine Education, *Conference Proceedings of Marine Education & Marine Technology,* 30 November – 1 December, Athens, pp.177-190, www.elint.org.gr.

Pare-Blagoev, J. 2007. The neural correlates of reading disorder: functional magnetic resonance imaging. In K.W. Fischer, J. H. Bernstein and M. h. Immordino-Yang (Eds), *Mind, Brain, and Education in Reading Disorders*. Cambridge:Cambridge University Press, 148-167.

Patton, M. Q. 1990. *Qualitative Evaluation and Research Methods*. CA:Sage Publications.

Picard, P. 1997. *Affective Computing*, MIT Press, USA:Cambridge MA.

Retalis, S. (eds.), 2005. Educational Technology. The advanced internet technologies in learning service, Athens:Kastaniotis Editions (in greek).

Tsoukalas, V. Papachristos, D. Mattheu, E. Tsoumas, N. 2008. Marine Engineers' Training: Educational Assessment of Engine Room Simulators, *WMU Journal of Maritime Affairs,* Vol.7, No.2, pp.429-448, ISSN 1651-436X, Current Awareness Bulletin, Vol. XX-No.10, Dec. 2008, IMO Maritime Knowledge Centre, pp.7.

Photo 1. Biometric tool (web camera & software 'Face Analysis') in action (back side).

Photo 2. Biometric tool (web camera & software 'Face Analysis') in action (side face).

14. Methodological Approaches to the Design of Business Games and Definition of Marine Specialists Training Content

S. Moyseenko & L. Meyler
Baltic Fishing Fleet State Academy, Kaliningrad, Russia

ABSTRACT: Knowledge owned by marine specialists demand only in rare cases of extreme navigation situations. In this regard, the development of training game methods for actions in such situations is a very actual problem. The complex of professional business games is elaborated in the Baltic Fishing Fleet State Academy (BFFSA). The scenario of the business game is being developed in line with given objectives. For example, the main goal of the business game "Safety ensuring of shipping" is formation of abilities and skills for management processes of safety ensuring in the field of cargo transportation.

1 INTRODUCTION

The article discusses the methodological approaches to the development of conceptual designs of professional business games for seafarers, and proposes a method for determining the content of training and development of the specialists' professional competence both directly during games and after game analysis. Professional development of marine specialists supposes learning new theoretical knowledge and its actualization. However, in actual practice of navigation, many theoretical developments are not often in demand, and for this reason, a specialist not familiar with the relevant skills in the proper degree[1].

Such a gap can be eliminated through "artificial / virtual" methods of knowledge actualization: training and business games. The effectiveness of these methods is proved into practice. But, if various kinds of training (for example, applying radar stations for ships passing, fire fighting, etc.) are used rather widely, business games didn't receive proper dissemination. The development of business games for maritime specialists requires high professionalism both from the developers themselves, and from teachers / experts, who lead the game and generate situations in the play activity of trainees.

2 PURPOSE AND TASKS OF THE BUSINESS GAME

Let's consider some methodological approaches to business games designing and usage of gaming methods to determine the content of training / development of maritime specialists' professionalism.

The main purpose of the business game is formation of abilities and skills of management of the processes of the cargo transportation safety[2,3].

The following tasks are necessary to solve in order to achieve this purpose:

- determining professional preparedness of specialists to act in difficult circumstances and unusual situations;
- developing skills of analysis and decision-making during process of professional activity;
- study of integrative processes of marine engineer's professional preparedness formation;
- analysis of maritime safety ensuring problems and efficiency of fleet commercial operation.

3 THE BUSINESS GAME DESIGNING

Preliminary design of the business game "Ensuring the safety of maritime cargo transportation" is given in the Table 1.

Table 1. Scenario plan of the business (imitating) game "Ensuring the safety of maritime cargo transportation"

Plot	Episodes	Actions
1. Preparation for the game	1.1. Instructing the game organizers and experts. 1.2. Instructing the game participants, studying of regulations, systems of the game control and evaluation of participants actions. 1.3.Determination of play groups and roles distribution.	Theme, goals, tasks, area of responsibility, regulations. Instructing, distribution and study of materials, etc. Clarification of the game governance, penalties and bonuses systems, etc. Formation of play groups and roles distribution.
2. Incoming control of participants' knowledge and skills	2.1. A participant's testing. 2.2. Definition of testing results.	Presentation of a test card to a participant of the game. Explanation the task of testing and an evaluation procedure.
3. A ship voyage planning	3.1. Definition of initial information, its collection and analysis. 3.2. Route selection of the voyage to the port of loading/unloading. 3.3. Project and the ship cargo plan development according to the criteria of safety and effectiveness. 3.4. Planning of ship stores, material and technical supply, crew completing and training for the voyage. 3.5. Ship voyage planning. 3.6. Planning safety actions.	A ship owner sends the ship's captain a telex containing instructions on the next voyage: ports of the ship loading/unloading, kind, type and quantity of cargo, special cargo characteristics, time for loading, loading/unloading rates, handling procedures of shipping documents, laytime calculation, etc. The captain of the ship distributes tasks to prepare the ship for the voyage.
4. Loading the ship	4.1. Coordination a cargo plan with stevedores and the action plan for work safety providing and environmental protection. 4.2. Monitoring the condition status of cargo, its stowage in holds and storage, mounting, etc. 4.3. Control of hydrometeorological conditions. 4.4. Drawing up primary shipping documents. 4.5. Drawing up protest letters in case of infraction of rules by stevedores in loading or delivery of cargo which has "defects ", as well as of the ship structures	The head of the game and experts performing the roles of shippers, agents, inspectors, surveyors, stevedores, etc. generate introductory data oriented to the growing complexity of the situation. Situation of a conflict between the ship and the shipper regarding the quality of packaging and labeling cargo is created. The situation of the ship structures or cargo damage due to negligence of dockers (crane damage during loading operators) is initiated. Participants of the game execute the required documents (notices, protests and acts of damage, etc.) and try to find compromise solutions. Fixing knowledge and skills needed for solving professional problems and situations, definition of "gaps" in knowledge and skills.
5. Issuing documents for cargo (Bills of Lading, cargo manifest, etc.)	5.1. Issuing Bills of Lading, addition of remarks and laytime commencement. 5.2. Drawing up permission for the ship to leave the port.	Experts initiate a conflict in terms of remarks including in the bill of lading, as well as in the act of laytime preparation. The situation when port authorities' claims concerning technical condition of the ship, etc. are simulated.
6. Preparation of the ship for leaving the port	6.1. Implementation of actions in accordance with the regulations for navigation safety. 6.2. Definition the time of the pilot's arrival onboard, the required number of tugs, a weather forecast, navigation conditions, etc. 6.3. Drawing up a statement of a fact the ship's readiness to a voyage.	Implementation of inspections in accordance with checklists. Diagnostics of technical means, etc. Discussing with the agent problems related to leave the ship at sea. Clarification the weather forecast and navigation conditions at the time of the ship leave. Filling in the logbook and other forms of documents. Experts initiate conflicts, complicate the situation. Simultaneous fixation of the captain's and his mate's errors as well as errors of experts.
7. Analysis and summarizing	7.1. Organization of discussing the game results. of the game. 7.2. Estimation of the working groups' actions in the game. 7.3. Formation the block of "gaps" in the knowledge and skills of the players, typical mistakes and miscalculations. 7.4. Final selection of learning content and development of professionalism of marine	Reports of the play groups leaders, self-evaluation of the actions of the game participants with an emphasis the attention on the occurred errors and the "gaps" in knowledge and skills, relation to the game and wishes. The experimental material accumulated during the game is systematized and processed in order it will be possible to carry out the procedure of training content selection, to build the system of

	specialists. 7.5. Specification of the training programs for specialists for the postgame period. 7.6. Consulting assistance to the participants of the game in development of their programs of self-development/self-designing. 7.7. Development by the game organizers of the postgame activity program concerning the game and preparation of the report on the game simulation experiment	subject knowledge and ways of their integration. Training programs are corrected according to the results of the game experiment if participants continue training in play groups. Players receive consulting assistance, including assistance in development of the program of self-training and self-development if they continue studies independently.
8. Postgame activity concerning the game.	8.1. Processing the results of the game simulation experiment with a goal to determine dependences, rational methods of formation of the professional personality, knowledge integration and configurators building, etc. 8.2. Conducting methodological seminars for teachers and experts. 8.3. Development the complex of purpose-oriented programs of vocational development of marine specialists and ways to adapt these programs to a person.	Results of the experiment are processed taking into account the earlier obtained data from other experiments of the same direction, that allows to accumulate the empirical material. Comparing the results, establishing dependences, confirmation of the previous results is considerable contribution to the development of our theoretical representations about the subject of the study.

4 THE BUSINESS GAME CONDUCTING

The game provides simulating six directly linked to the game plots.

Participants get instructions and learn the game rules as a part of the first plot. At that time the head of the game, formulating goals of the game simulation experiment and analyzing the situation, notes existing substantial contradictions between the desirable and realizable, and thereby "running" problematization processes, goals setting and self-determinations.

The second plot provides preliminary estimates of participants' readiness to the game. In case of unsatisfactory test results it is assumed that such a result helps to intensify the process of motivation of specialists for developing professionalism. "Gaps" in knowledge and skills found as the test results are eliminated by the head of the game decision. Thus, the compensatory function of education is realized, i.e. advices can be given to participants of the game and special literature for self-study may be recommended.

The third, fourth and sixth plots are extremely important because at these stages in reality many of the major issues of navigation safety ensuring and sea cargo transportation are solved. It is assumed that not only typical real situations are fulfilled at these stages, but complex non-standard situations that happen rarely in real activity are simulated too, because serious negative consequences can entail, in cases of such situations appearance, if adequate solutions will not be found by responsible specialists.

Thus, the task of the game head and experts playing roles of officials, who under certain conditions, can counteract the captain of the ship and other persons involved in the process of sea cargo transportation is to generate episodes and situations "provoking" a conflict and thereby substantially complicating making a solution of professional tasks. At the same time, as "home prepared" tasks, as actions according to a present situation can be performed for generating game situations. It allows to implement the game variation into the real game situation, i.e. the structure of the game has many degrees of freedom. Thus, the game adapts depending on the goals and specific problems requiring the solution.

Game participants fix results of the analysis, calculations, decisions and the effect of these decisions. They take into account the specific conditions, professional experience of all participants of the game group. The experts record the work of groups and individuals in the each episode and plot. Their duties include clear fixating of errors, "gaps" in knowledge, abilities and skills of specialists and their ability to integrate with various kinds of knowledge for solving complex professional problems. In addition experts observe the behavior of game participants in difficult situations.

During the game experts and the head of the game analyze the activities of participants and estimate their performance, identify areas of knowledge and skills in which it is useful to hold substantive and methodological consultations. For example, our experience in business games shows that practically always there is a need for methodological consultations in a systematic approach methodology, a system analysis, designing without prototypes [2], etc.

Decisions obtained for each plot are discussed by all participants of the game. Representatives of the play groups make reports, where the idea of the design decision is revealed, as well its motivation and implementation methods with evaluation of possible consequences of the decision implementation.

Speakers answer questions from experts and other participants of the game, fixing critical remarks and opponents' suggestions. Final estimates for each plot are determined after the public discussion of the play groups' reports.

All working materials relating to the analysis of situations in each plot, developing design solutions, decisions' motivation and choice of the decisions implementing methods are given to the game head for the further examination in accordance with the objectives of the game simulation experiment.

Processing game results requires certain time. Therefore, it is possible to realize partially selection of the content of the training and professional development directly during the game, as it was mentioned above. But careful analysis of the game materials is held in the postgame period. In the process of such an analysis it is often possible to detect some important regularity, to get an understanding of some of the integration processes, interdisciplinary system links, and to evaluate the effectiveness of various knowledge integrating methods for professional problems solving [3,4].

The experimental results allow to select the content of training and development of marine specialists on the higher qualitative level, as well as to find new plots of the game and new opportunities for the whole game.

5 THE TRAINING CONTENT SELECTION

An example of the training content selection and professional development of ships masters on the results of game simulating experiments is given in the Table 2. Matrix representation of the content selection scheme for marine specialists training and development allows to realize a deductive method of analysis of each plot, differentiation of activities and subject knowledge required for their implementation. Further, it is possible to determine what skills and abilities a specialist must have to solve professional problems and, therefore, what a specialist's skills and abilities it is necessary to develop to achieve a high level of professionalism.

Table 2. Matrix representation of the content selection scheme for marine specialists training and development

Plot or operation	Acts of activity	Subject knowledge	Skills and abilities
Ship's voyage planning.	Route selection of the voyage to the port of destination	Hydrometeorology and. oceanography Navigation and sailing directions. Cyclones and anticyclones, sea currents. Aid to navigation, etc.	Be able: to find sources of information; to analyze synoptic chart data and long-term forecasts; to determine risk factors and their assessment correlated to concrete constructive and exploitations characteristics of the ship, its purpose; to evaluate the impact of the navigation conditions on the operating parameters of the ship; to find necessary documents, regulating cargo transportation, to apply them; to make the cargo plan showing the holds rotation during loading; to perform calculations of stowage; to calculate ship stability and sitting, bending moments and shear forces; to assess the impact of the extreme conditions on the longitudinal and local strength of the ship, etc.
Planning measures to ensure the safety of marine cargo Transportation	The project and plan of ship loading development.	The theory of the ship. Theoretical mechanics. Strength of materials.Regulations for the cargo transportation Requirements of the international convention (SOLAS, MARPOL), on load line mark, etc. Ship's stability at high angles of heel, local and general ship's longitudinal strength, ship's unsinkability. Requirements and recommended schemes of different cargo strap-ping. International Sea Law. International regulations related to navigation. Rules of technical exploitation. Fire safety rules. Navigation safety rules. Knowledge of rules and their application, knowledge of basic legal acts (territorial waters, economic zones, etc.)	Be able: to find necessary document, to assess it's adequacy to the studied problem; to apply creatively legal acts, regulations, recommendations in order to solve practical problems; to present systemically whole range of measures to ensure to present systemically whole range of measures to ensure the safety of maritime cargo transportation and to correlate everything with methods and means of these measures practical implementation. Fixation performance fact in the logbook.

Plot or operation	Acts of activity	Subject knowledge	Skills and abilities
Execution of shipping documents	Issuing the Bill of Lading	Commercial work in the merchant fleet, a notion of the Bill of Lading, its kinds, and functions; master's remarks to the Bill of Lading concerning cargo quality or it's packing, etc.	Be able: to issue the Bill of Lading correctly and formulate remarks (if there are any) in accordance with cargo insurer's recommendation; to champion / to protect the comercial interests of the ship owner or the charterer in the case of conflict with shippers; to formulate claims and to justify them; to prove invalidity of the claims to the ship.

6 CONCLUSIONS

Efficient usage of gaming methods for training and selection of training content is determined primarily by the fact that, if the game participants cannot find solution to any given situation, then the following processes are started:

1 self-disqualification, i.e. a game participant discover himself a lack of knowledge and became aware of the need of its replenishment;
2 motivation of new knowledge assimilation and further professional development;
3 selection of the content of personal-oriented compensatory and developed education;
4 development of new game plots and game trajectory correcting;
5 development of personal-oriented trajectories of specialists self-development and the organizational - pedagogical conditions of their realization.

REFERENCES

Moyseenko S., Socio-pedagogical conditions for continuing professional education of marine engineers, Monograph, BFFSA, Kaliningrad 2004.

Moyseenko S., Meyler L., Theoretical and practical problems of specialists' professional competence development in the field of maritime transport organization, Joint International IGIP-SEFI Annual Conference, Trnava 2010.

Meyler L., Moyseenko S., Development Prospects for the Maritime Transport Complex of the Kaliningrad Region and Professional Training, „Technical Coperation in Maritime Education and Training", Proceedings of the 11-th Annual General Assembly International Association of Maritime Universities, Pusan 2010.

Pidkasistiy P., Game technology in learning and development, Russian pedagogical agency, Moscow 1996.

Wells, R. A., "Management Games and Simulations in Management Development: An Introduction", *Journal of Management Development,* 1990, **9** (2)

Platov V., Business games: development, organization, conducting, Nauka, Moscow 1991.

15. A Door Opener: Teaching Cross Cultural Competence to Seafarers

C. Chirea-Ungureanu
Constanta Maritime University, Constanta, Romania

P.-E. Rosenhave
Vestfold University College, Tonsberg, Norway

ABSTRACT: The importance of developing cultural competence in maritime professionals is increasingly being recognized. Seafarers seek knowledge to help them cope with the growing diversity of their employers, leaders and colleagues. However, even though requirements designed to address cultural competence are incorporated into maritime school curricula, the institutional culture of maritime education systematically tends to foster static and essentialist conceptions of "culture" as applied to seafarers. Many questions emerge when we try to teach in a way that brings alive the humanity of mariners. These questions are waiting for their answers, so in our paper we shall try to find and explain some approaches and ways of teaching and research as the goal is to provide maritime professionals with practical wisdom in comprehending what is the seafarers' life on board ship.

1 CROSS CULTURAL COMMUNICATION IS A 21ST CENTURY SKILL

In the 21st Century, people need to have the ability to get along with other cultures, ethnic groups, and races at all levels of society. In a diverse society, there is a pressing need to communicate cross-culturally in and out of the classroom. As the planet ages and communities become more multilingual, classrooms reflect a global society where people must learn to interact and create harmony.

Teachers must be master communicators who can influence young minds in positive ways. Learning how to instruct students in the art of cross-cultural communication is a necessary goal of effective educators. More importantly, teachers must take the lead and develop strategies that assure their students will learn not only navigation and maritime technology, but also cross-cultural communication skills.

The difficulty is that no-one is "born great" at communicating with others, because:

1 Parents never taught their children about effective communication (probably because no one taught them.)
2 In school effective communication is not generally taught: children have to sit in their seats, be quiet, and raise their hands to speak and to recite facts upon demand.
3 In the workplace, there might be some mandatory training about effective communication, but it is known that information about "effective communication" in the workplace can be devastating to interactions.

When communication in the workplace is taught, it is usually explained from one of two perspectives:

1 To illustrate how to communicate to influence someone or
2 To illustrate how to communicate with people to be more efficient and get more work done.

Communicating with the intention to influence or communicating with the intention to be efficient or effective may lead to more sales and higher productivity in the workplace. But, when attempting to use either of those two strategies for onboard communication, it becomes quickly apparent how easy it is to destroy it in almost no time at all.

Good communication isn't created by efficiency or influence. It is created by connection, interaction, balance and understanding.

Communicating one's ideas is the key to knowledge. As such, it is extremely important for educators to elicit academic performance from students that is based on communication skills.

There are three fundamental elements which embody the spirit of cross-cultural communication:

- Intercultural awareness
- Intercultural sensitivity
- Intercultural communication competence (Cole.C.W, Prichard B. &Trenkner P.2005).

2 BUILDING STUDENTS' AWARENESS AND SENSITIVITY IS FUNDAMENTAL

As a teacher, one must incorporate each element into lesson planning. For instance, through development of intercultural awareness, students learn to identify and accept cultural similarities and differences. Some methods of instruction that improve intercultural awareness are: reading assignments, and watching drama. In terms of intercultural sensitivity, students must learn to respect and tolerate cultural differences of their peers. Being able to walk in another person's shoes is an acquired ability that takes training and practice. Methods of instruction that enhance intercultural sensitivity are role-laying, group discussions, and paired exercises (Littrell et al.2005).

The importance of developing cultural competence in maritime professionals is increasingly being recognized. Seafarers seek knowledge to help them cope with the growing diversity of their employers, leaders and colleagues. However, even though requirements designed to address cultural competence are incorporated into maritime curricula, the institutional culture of maritime education systematically tends to foster static and essentialist conceptions of "culture" as applied to seafarers.

So what is the best way to give nautical students a more flexible and useful knowledge of culture to work effectively with multilingual crewmembers? In a short amount of time it may seem that the best method is to explain some basic cultural characteristics to look for and use in maritime encounters with multilingual seafarers. However, while helpful in providing some guidelines to work with, this approach stereotypes and objectifies seafarers by ignoring individual variation and the fluidity of cultural change. It creates resistance in students who feel they are not part of the discussion. This method of teaching is distancing by its very nature, as it describes a constructed group rather than individual mariners, which is what we actually encounter in practice.

3 COMMUNICATION COMPETENCE IMPROVES STUDENTS' TOLERANCE LEVELS

Intercultural communication competence is the major goal of students who develop both intercultural awareness and sensitivity. Communication competence reflects having the ability to negotiate and interact well across cultures. Reading, writing, and speaking are methods of instruction that help to increase intercultural communication competence. By increasing the level of discourse in the classroom, a teacher can expect students to make cultural connections that may last forever.

It is an educator's responsibility to ensure that his or her classroom supports intercultural awareness, sensitivity and communication competence. Without an understanding of cultural diversity it is possible for teachers to neglect the different needs of every student. Developing curriculum that addresses cross-communication is one solution to this ever present problem. Intercultural Education aims to go beyond passive coexistence, to achieve a developing and sustainable way of living together in multilingual societies through the creation of *understanding* of, *respect* for and *dialogue* between the different cultural groups onboard ship.

Intercultural education onboard cannot be just a simple 'add on' to the regular nautical curriculum. It needs to concern the learning environment as a whole, as well as other dimensions of educational processes, such as academic life and decision making, teacher education and training, curricula, languages of instruction, teaching methods and student interactions, and learning materials (UNESCO 2003).

Many questions emerge when attempting to teach in a way that brings alive the humanity of mariners. How can students hold general knowledge without it overwhelming their perceptions? How can they remain open to learning from marines who went before them onboard ship? How can they maintain awareness that any mariner is like all other humans in some ways, like some other humans in certain ways, and also has a particular life story?

One approach is to use narrative, by emphasizing how telling, receiving, and creating stories are integral features of maritime practice, teaching, and research. The goal is to provide maritime professionals with practical wisdom in comprehending what is the seafarers' life on board ship. Narrative humanises by putting the mariner first and the cultural group second. A narrative approach helps students make connections, see similarities as well as differences, and deal with complexity rather than reduce to simplicity.

Asking people to deal with complexity, when they want simplicity is a struggle. It challenges nautical students to deal with vulnerability when they seek certainty and humility when they seek competence. Our experience shows that many students are up to the challenge and that what we discuss through narrative may actually prove to be more useful and more immediately practical in terms of everyday maritime experience than a detailed list of general cultural characteristics.

In cross-cultural training and living within a multilingual environment, the goal of the seafarer is to learn about himself and others. Just as the desire to learn another language arises from the desire to communicate with local people and understand the new world, the seafarer also will want to learn the silent language of cultures— his own and his host onboard environment. In trying to appreciate the differences between his own culture and that on board ship, the seafarer may feel that he is supposed to like and accept all these differences. Cultural sensitivity, however, means knowing about and respecting the norms of the onboard culture, not necessarily liking them. The seafarer may, in fact, be frustrated or even offended by certain acts. In some cases, increased understanding will lead to greater respect, tolerance, and acceptance; in others, it just leads to enhanced awareness. The goal in cross-cultural training is to increase understanding, to equip the seafarer with a powerful set of skills, a framework to make sense of whatever he does and experiences as a seafarer so that he will be able to interact successfully with the multilingual environment. Whilst often understanding much of what has been happening, many actions, attitudes, values—entire ways of thinking and behaving— may on occasion surprise, puzzle, or even shock the seafarer. On the other hand, the latter may also be unaware of what he has in common with other multilingual crewmembers. People in any culture, for example, need to find an acceptable way to express anger, cope with sadness, manage conflict, show respect, demonstrate love, or deal with sexuality. When examining the differences between two cultures, one often looks at different ways of answering the same questions. If the similarities are not clear, it is because the ways of acting or thinking differently are what produced the most challenge and tension. What people have in *common* often goes unnoticed, but it is one of the important parts of life onboard ship.

Keep in mind, too, that culture is just one of numerous influences on behaviour. People can differ from each other in many other aspects as well. Could the miscommunication or misunderstanding between two seafarers of different nationalities actually be the result of a difference in job position, personality, age, generation, or gender, and not a cultural difference? In trying to understand the role culture plays in behaviour, it should be noted that personal differences often play as great or even a greater role.

It is important to understand that what people do and say in a particular culture, whether it is yours own or that of a host onboard environment, are not arbitrary and spontaneous, but are consistent with what people in that culture value and believe in. By knowing people's values and beliefs, it is possible to anticipate and predict their behaviour. Once a seafarer is no longer caught off guard by the actions of host onboard crew members and once he does not simply react to these, the seafarer is well on his way to successful cultural adjustment. Moreover, once the seafarer comes to accept that people behave the way they do for a reason, whatever he may think of that reason, he can go beyond simply reacting to that behaviour and figure out how to work with it. Knowing where host onboard behaviour is coming from doesn't mean that the seafarer has to like or accept it, but it should mean that he is no longer surprised by it—and that is a considerable step toward successful interaction.

Designing the right lesson plans is not enough. Teachers must use the plans consistently and make sure that students understand learner objectives. Doing this they will ensure that students are focused on academic success, as they gradually develop the capacity to tolerate others' differences.

The intercultural competence is required not only in interactions between people and groups, but in ethnic and international relations, where different cultures may interfere. That is why the education aims gradually to build the needed intercultural skills, aiming to train for objectivity in dealing with other cultures and their representatives.

The general model of curriculum design involves the following steps, performed in the following order:

1 *What shall I do?* This step implies the targets formula (Establishing of the general aim of educational program based on the beneficiary's needs)
2 *What shall I use?* This step implies the providing resources (Appropriate core objectives) and restriction analysis (time, learning abilities etc.)
3 *How shall I do it?* This step implies the working strategy (Appropriate learning tasks and situations consistent with the objectives)
4 *How shall I know that I have done the right thing?* This step implies the development of assessment tools.

In their review of the cross-cultural training literature, Littrell *et al* (2006) have identified six approaches to the delivery of intercultural training programs:

Attribution Training – The aim of attribution training is for the trainee to interpret behaviour from the viewpoint of the host culture nationals.

Culture Awareness Training – This approach uses T-groups (cultural sensitivity training groups) to guide the exploration of the trainee's culture of origin. This entails delving into cultural biases and values, based on the premise that a deep awareness of the trainee's own culture will lead to a better understanding of the dynamics of intercultural communication.

Interaction Training – The trainee employee benefits from on-the-job training, learning the ropes from a former trainee who is already performing the job function.

Language Training – Language acquisition is an important element of adjustment to a new cultural environment. While fluency is always the goal of a language training program, making the effort to speak even simple phrases in the local language generates enormous goodwill among host nationals.

Didactic Training – The goal of this fact-based training is to supply practical information to the trainee regarding living conditions, cultural differences, job details, and other requirements for establishing a lifestyle in the new locale. "In addition," write Littrell *et al*, it provides "a framework for evaluating new situations that will be encountered [and]... enhances the cognitive skills that enable the trainee to understand the host culture."

Didactic training is delivered via any combination of informal briefings, written materials, lectures, and cultural assimilators. The latter is a training tool that allows the trainee to consider how best to respond to various authentic cross-cultural situations through the use of critical incidents. According to the authors, it has been established that the cultural assimilator lessens the incidence of adjustment problems.

Experiential training – This approach develops intercultural communication skills through techniques such as simulations, and role-plays. As its name implies, it involves learning by doing. The most effective training approaches are those that incorporate experiential learning techniques.

The training approach had several advantages:

- It moved the focus from the trainer to the trainee.
- It compelled trainees to take responsibility for their own learning.
- It stressed problem-solving rather than memorization of facts.

It put the emphasis on learning how to learn.

This idea of learning how to learn is still an important theme, as cross cultural training cannot prepare seafarers for every possible situation likely to be encountered in the host onboard environment. Essentially, the seafarer is taught how to learn and acquire information about another culture.

The majority of multinational shipping companies providing intercultural training to their employees do so through informal briefings. The field is still developing, however, and new advances are emerging that may someday change the face of cross-cultural training.

Cross-cultural training improves skills that lead to seafarer psychological comfort, including intercultural competence and effective interpersonal communication. It comprises three dimensions: work adjustment, interaction adjustment with host onboard environment, and general adjustment to the foreign culture. Although many factors affect overall seafarer adjustment in a new onboard environment, numerous studies have suggested that cross-cultural training can contribute significantly to adjustment in each of these dimensions. While cross cultural training alone cannot guarantee successful adjustment to a novel culture, the studies suggest that relevant, honest, and current training content generates more realistic expectations about life in the new onboard environment. They found that cross-cultural training was positively related to:

- self-development and self-confidence;
- the establishment of personal relationship with host onboard environment;
- overall feelings of well-being and satisfaction; and
- cognitive skills development with regard to perceptions of host onboard environment.

It is the absence of this connection between the self and the new cultural onboard environment that leads to what Kim in her book *Becoming Intercultural: An Integrative Theory of Communication and Cross-Cultural Adaptation*, calls "a serious disequilibrium within the stranger's psyche." It can manifest itself in the following symptoms:

- Sadness
- Loneliness
- Homesickness
- Idealizing the home culture
- Stereotyping host culture nationals
- Dissatisfaction with life in general
- Loss of sense of humour
- Sense of isolation, withdrawal from society
- Overwhelming and irrational fears related to the host country
- Irritability, resentment
- Family conflict
- Loss of identity
- Feelings of inadequacy or insecurity
- Negative self-image
- Developing obsessions (health, cleanliness)
- Cognitive fogginess, lack of concentration
- Depression

People interact! Therefore the mere existence of some kind of cross-cultural training is not sufficient. It is recommended that the training be individually designed to accommodate the particular situation as outlined above. Cultural distance – the extent to which two cultures are similar or different – should also be taken into account. The greater the cultural distance between the home and host onboard cultures, the more necessary cross-cultural training is.

It is clear that cross-cultural training creates favourable conditions for cross-cultural learning to occur. When it's relevant to the seafarer's situation, it makes possible the development of realistic expectations about life in the host onboard environment, and increases skills that lead to overall seafarer adjustment. Cross-cultural training provides realistic expectations, and insight into managing cultural differences.

The cultural composition of societies is today growing even more complex through increasing migratory movements from one country to another and from rural to urban regions. Whereas indigenous peoples and other minority groups can look back on a long historical tradition in a given region, today's migratory movements tend to produce culturally fragmented, usually urban or semi-urban societies, which present specific challenges for educational policies (UNESCO 2001).

The distinct aims of Intercultural Education can be summarized under the headings of 'the four pillars of education' as identified by the International Commission on Education for the Twenty-First Century (Delors, 1996). According to the conclusions of the Commission, education should be broadly based on the pillars of:

1. Learning to know, by "*combining sufficiently broad general knowledge with the opportunity to work in-depth on a small number of projects*" (Delors, 1996). The Commission states further, "*a general education brings a person into contact with other languages and areas of knowledge, and... makes communication possible*" (Delors, 1996). These results of a general education represent some of the fundamental skills to be transmitted through intercultural education.

2. Learning to do, in order to "*acquire not only an occupational skill but also, more broadly, the competence to deal with many situations and to work in teams*" (Delors, 1996). In the national and international context, learning to do also includes the acquisition of necessary competencies that enable the individual to find a place in society.

3. Learning to live together, by "*developing an understanding of other people and an appreciation of interdependence – carrying out joint projects and learning to manage conflicts – in a spirit of respect for the values of pluralism, mutual understanding... peace*" (Delors, 1996) and cultural diversity. In short, the learner needs to acquire knowledge, skills and values that contribute to a spirit of solidarity and co-operation among diverse individuals and groups in society.

4. Learning to be, "*so as to better develop one's personality and be able to act with ever greater autonomy, judgment and personal responsibility. In that respect, education must not disregard any aspect of a person's potential...*" (Delors,1996) such as his or her cultural potential, and it must be based on the right to difference. These values strengthen a sense of identity and personal meaning for the learner, as well as benefiting of their cognitive capacity.

Several studies have examined the problems and potential solutions when facing an intercultural environment at work, but on a ship an additional dimension is added. Not only do the seafarers have to ensure good communication during working hours. The ship is also a learning environment and a social environment, where people eat and live together, often for long periods on end. For this reason Intercultural communication is what makes the teamwork function on a ship. It gives you a positive social environment, fewer problems and most certainly fewer accidents.

An extreme example: the bulk carrier Bright Field, which ran into a shopping complex in New Orleans in 1996, leaving 66 people injured, illustrates an extreme situation with a crew and a pilot from different cultures: American and Chinese. The word "no" is a very impolite word to the Chinese especially to an authority such as a pilot. Since the pilot was not able to understand the communication in Chinese between the engine room and the bridge, he was left unaware of the engine problems and could take no preventive action to mitigate the accident.

It is no doubt difficult for seafarers that communicate in their native languages and perhaps simplified English in their day-to-day communication to suddenly muster a good command of a standard marine vocabulary according to the STCW convention, when an emergency situation occurs.

4 CONCLUSIONS

Providing realistic expectations of life in the new locale, and the skills to deal with intercultural interactions, should therefore reduce the stress and ambiguity seafarers experience when dealing with the unknown onboard culture, thus improving adjustment. However, studies on the effectiveness of cross-cultural training have produced mixed results, perhaps because there is no consensus on what, exactly, it entails.

Education systems need to be responsive to the specific educational needs of all minorities, including migrants and indigenous peoples. Among the issues to be considered is how to foster the cultural, social and economic vitality of such communities through effective and adequate educational programmes that are based on the cultural perspectives and orientations of the learners, while at the same time providing for the acquisition of knowledge and skills that enable them to participate fully in the larger society.

Improved crew communication through training and education can reduce the risk of accidents as long as it is based on fundamental knowledge of the dynamics of crew interaction and communication. Leadership onboard necessitates cross- cultural competency to revoke cultural differences in order to get the best out of a multicultural team. What you can do is decide, if you want to be limiting or non-limiting in your communication, listening or non- listening. You have the choice to open up professional communication.

REFERENCES

Cole, C.W., Pritchard, B, & Trenkner, P. (2005) "The professional profile of Maritime English instructor (PROFS): an interim report" in *Maritime Security and MET, Proceedings of the International Association of Maritime Universities (IAMU) Sixth Annual General Assembly and Conference*, 65-71. Southampton: WIT Press.

Delors, J., (1996) "*Learning: The Treasure Within – Report to UNESCO of the International Commission on Education for the Twenty-first Century"*, UNESCO, pp: 87-97.
Kim, Young Yun 2001. *Becoming Intercultural: An Integrative Theory of Communication and Cross-Cultural Adaptation.* Thousand Oaks, CA: Sage Publications Inc.
Kim, Chai. *Improving Intercultural Communication Skills: A Challenge facing Institutions of Higher Education in the 21st Century.* (Accessed December 6, 2010).
Littrell, Lisa N., Salas, Eduardo, et al. 2006. Expatriate Preparation: A Critical Analysis of 25 Years of Cross-Cultural Training Research. *Human Resource Development Review* 5:3 (2006): 355-388.
UNESCO (1992): *International Conference on Edu- cation, 43rd Session, The Contribution of Education to Cultural Development*, p.5, §10.
UNESCO (2001): *Universal Declaration on Cultural Diversity (*Culture is "*the whole complex of distinctive spiritual, material, intellectual and emotional features that characterize a society or social group. It includes not only the arts and letters, but also modes of life, the fundamental rights of the human being, value systems, traditions and beliefs*")
UNESCO (2003): *Education in a Multilingual World*, UNESCO Education Position Paper (It discusses the use of mother tongue (or first language), as language of instruction for initial instruction and literacy, the importance of bilingual or multilingual education (i.e. the use of more than one language of instruction), and language teaching with a strong cultural component).

16. Considerations on Maritime Watch Keeping Officers' Vocational Training

L. C. Stan & N. Buzbuchi
Constanta Maritime University, Constanta, Romania

ABSTRACT: The activities on the board of the ships are based on competences and skills. In order to have competent people, you must to create them. This is the role of the maritime training system, to create competent persons for the maritime field. Part of this system is the vocational training for the deck officers. Maybe the most important role of the vocational system is to create competences based on the previous skills and knowledge acquired during the practice period on the board of the ship.

1 INTRODUCTION

The actual Constanta Maritime University maritime academic training is based on the Bologna process. Inside of this process, the training period has four years and at the end, the graduates receive a double qualification, as officers for maritime industry and as engineers for complementary industries. To achieve this double qualification is necessary, during the four years of study, to cover the special curricula for both of them. The curricula for maritime competencies are based on the STCW Convention requirements and for the engineer qualification is structured based on the national requirements.

These double qualifications are very useful for those who do not want to have a long career on sea, offering them the possibility to find a job on shore after a number of years on sea, or from the beginning, after the graduation for those who do not want to work on the sea. On the other hand, this system is not convenient for a person with years of practice on sea as seaman or motorman and who wants to increase the level of training and pass to a superior rank, as officer. In this case, more efficient will be to have implemented a vocational system of training, shorter and based exclusively on the STCW Convention requirements for duty onboard ships as watch keeping officer.

2 VOCATIONAL TRAINING INSIDE OF THE MARITIME ACADEMIC TRAINING SYSTEM

Starting from idea that the maritime vocational training is designed especially for those persons with a background in the maritime field, as ordinary seaman or helmsman, the training process must cover only the areas of interest which are not covered by the onboard activities.

This concept leads to the present stage of vocational maritime training that exists all over the world, based on two years of study, when the students learn only the curricula according with the STCW Convention requirements. During these two years, all necessary knowledge required for a duty activity on the ship bridge is compulsory to achieve. Taking into account, that the background knowledge, practical part, about these activities, already exists in the trainee, the main focus has to be on the theoretical knowledge and in this direction, more efficient, could be a modular system of training.

Inside of this modular system can be included modules about navigation techniques, ship handling, cargo handling, regulations, maritime English and other valuable knowledge for a safety activity on the bridge.

The modular system represents in the same time the solution for compression of three or four years of study in only two, making possible the coverage of all specialized curricula for the deck officer.

Another opportunity in the way of decrease the study period is represented by the possibility to exclude from the training period the on board practice, this part being considered covered during previous periods of working on board ships.

The main problem regarding the vocational maritime training is about the equivalence between Bologna process training, of four years, and the vocational training, of only two years. Due to the shorter period of training inside of the vocational system, the modular system will be difficult to be considered as academic level studies and to possess the same statute. Some actual opinions consider as optimum to accept the vocational system as training for the operational level and the long form, as training for the operational and managerial level too.

These opinions are based on the consideration that it is not possible to cover all the requested number of training hours for the managerial level in only two years of study.

Below, there are described an education scheme and an assessment system for the maritime deck officer's vocational training starting from the previous consideration and according with the present situation in Romanian maritime training system.

3 MARITIME DECK OFFICER'S VOCATIONAL EDUCATION SCHEME AND ASSESSMENT SYSTEM

The education scheme for the vocational training of a deck officer must be centred on the STCW Convention requirements and on the functions as navigation, cargo handling and stowage, controlling operations and care for the persons on board.

Building on these functions as modular system, the result will be a number of eight modules containing knowledge in the navigation field, four modules for the cargo handling, two modules for the controlling operations and care for the persons on board, two modules will include the complementary knowledge as electronics, marine engine and on board equipments other than navigation ones.

The navigation modules are:

- Module 1: Coastal navigation and celestial navigation
- Module 2: Navigation equipments and ship handling
- Module 3: Radar and electronic navigation
- Module 4: Voyage planning, Bridge Team Management, Collision Regulation

The modules about controlling operations and care for the persons on board are:

- Operation and maintenance of the ship
- Search and Rescue Operation

Also, adjacent to these principal modules will be other two modules dedicated exclusively to the compulsory IMO Courses, grouped in the basic and advanced courses.

- This structure will be distributed on four semesters as:
- First semester: Coastal Navigation, part of complementary courses as electronics and on board equipments other than navigation ones and basic IMO courses;
- Second semester: Coastal Navigation, Celestial Navigation, Navigation Equipments, Ship Handling, Operation and maintenance of the ship, Cargo handling and stowage and marine engine;
- Third semester: Radar Navigation, Electronic navigation, Collision Regulation, Cargo handling and stowage, Search and Rescue Operation;
- Fourth semester: Radar Navigation, Voyage Planning, Bridge Team Management, including Human Factor and advanced IMO Courses.

A semester of 16 to 18 weeks long is considered adequate in order to comprise the aforementioned structure and to cover the complete number of training hours requested by the STCW Convention.

The arrangement of the modules during the semesters permits to structure the knowledge in a natural order and thus, to have continuity.

A part of the modules can be arranged in another chronology, but it is very important to assure that the necessary knowledge is offered by a previous module in order to pass to a new one. So, in this context, is unnatural to place Celestial Navigation or Radar Navigation before Coastal Navigation, course which provides the basic knowledge for all the following navigation modules.

The introduction of the IMO courses is necessary in order to proceed to the certification as deck officer at the end of the training period.

During the courses development, it must be taken into consideration that the trainees have as background a period of practice on the ship board and some knowledge is possible to be known and in this way is necessary to see first how and what they know and to try to involve them actively in the training process and to facilitate the knowledge acquiring and understanding.

An important idea that has not be forgotten is that one of the main scopes of the vocational training in the maritime field is to create competences, competences that can be based on the previous skills developed during the periods of service on sea.

Any competence based system must be managed within recognized arrangements for assuring quality. The requirements for a quality standards system are included in the revised STCW Convention (Regulation I/8). They cover all training courses and programmes, examinations and assessments and the qualifications and experience of instructors and assessors. The regulation I/6 requires all training and assessments to be structured in accordance with documented programmes and procedures necessary to achieve the prescribed standard of competence and conducted and supervised by persons qualified in accordance with the convention.

It is absolutely necessary to define the standards of competence required, so that the assessor can make a judgment against those standards. Otherwise, each assessor will be guided by opinion and the own experience. Within the definition of these standards must be a clear indication as to the level of competence required.

Assessment is the process of obtaining and comparing evidence of competence with the standards. The sources of evidence could be:

- Direct observation: in-service experience, laboratory equipment training, simulation.
- Skills, proficiency and competency tests.
- Projects/assignments.
- Evidence from prior experience
- Questioning techniques: written, oral, by computer.

4 CONCLUSIONS

The activities on the board of the ships are based on competences and skills. In order to have competent people, you must to create them. This is the role of the maritime training system, to create competent persons for the maritime field. Part of this system is the vocational training for the deck officers. Maybe the most important role of the vocational system is to create competences based on the previous skills and knowledge acquired during the practice period on the board of the ship.

Also, the vocational training is a viable alternative for those individuals who have not enough time for a fourth years training, active individuals in the maritime field who want in the same time to increase their level of training and to have possibility to get an upper position on board, as officers.

The training scheme must satisfy in the same time the STCW Convention requirements and also to offer to the trainee the necessary knowledge for the future duties. In this way, it is considered as useful to use a modular scheme, containing a combination of traditional nautical sciences, as different types of navigation, cargo handling, ship handling, regulations and communication.

The assessment must offer the possibility to obtain and compare evidence of competence with the standards. The aim is to ensure that sufficient, reliable and verifiable evidences are available to enable an assessor to be satisfied that a candidate has the ability to work in accordance with the required standards.

The vocational training is in accordance with the International Maritime Organisation (IMO), STCW and shipping internationally recognized classification societies, as DNV, for training and bridge, engine and liquid cargo handling simulators but also Telenor Norway for additional equipment GOC - GMDSS. The training is accredited by National Council for Adult Vocational Training (CNFPA), Romanian Agency for Quality Assurance in Higher Education (ARACIS) and Romanian Naval Authority (RNA).

REFERENCES

Hanzu, R. & Stan, L., Grosan N., Varsami, A. 2009. *Particularities of cadets practice inside of a multinational crew*, 10th General Assambly of International Association of Maritime Universities, IAMU. St. Petersburg. Russia. published in MET trends in the XXI century. ISBN 978-5-9509-0046-4. Pub. Makarova. Russia; 99-105

Raicu, G. & Surugiu, F. & Stan, L. 2009. *Online teaching technique – The ground of life long learning concept*, 4th International Conference on Maritime Transport.Barcelona. Spain. ISBN 978-84-7653-891-3, pub. Universitat Politecnica de Catalunya. Spain: 507-516

Stan, L. *Reducing of maritime accidents caused by human factors using simulators in training process*, The Knowledge-Based Organization The 14th International Conference, "Nicolae Bălcescu" Land Forces Academy Publishing House, 28-30 November. Sibiu, Romania;

Stan, L. & Buzbuchi, N. 2009. *The importance of the educational factor to assure the safe and security on the sea*, Proceedings oh the 8th International Conference TransNav 2009 – Gdynia, Poland, published in Maritime navigation and safety of sea transportation, pag. 751-755, ISBN 978-0-415-80479-0. Gdynia, Poland;

Stan, L. & Hanzu-Pazara, R., Bocănete, P. 2009. *The training system and technology challenges,* Source: Proceedings of the 6th International Conference on Management Technological Changes, Book II, ISBN 978-960-89832-8-1. Alexandroupolis. Greece: 259-263

International Maritime Organization, Model Course 6.09 - Training Course For Instructors, IMO, London, 1991.

International Maritime Organization, Model Course 3.12 - Examination and Certification of Seafarers, IMO, London, 1992.

17. Simulation Training for Replenishment at Sea (RAS) Operations: Addressing the Unique Problems of 'Close-Alongside' and 'In-line' Support for Multi-Streamer Seismic Survey Vessels Underway

E. Doyle
Cork, Ireland

ABSTRACT: Modern siesmic survey vessels in 'production', may tow twelve or more streamers, each of which can be six to eight kilometres long. Together with associated paravanes, tail-buoys and acoustic 'guns', the streamer spread width of such wide-tow configurations can extend to 1200 metres. The physical deployment and recovery of such an extensive array is time-consuming and expensive. The entire survey operation requires the constant attendence of a suitable offshore support vessel (OSV) to act in the role of 'chase vessel', but more critically, to provide close replenishment support underway and, when required, rapid emergency towing assistance.

While naval crews rightly claim a near monopoly on the skills-set necessary for underway replenishment, the naval RAS exercise almost never involves the supply and receiving vessels engaging 'close-alongside'. The seismic/OSV replenishment operation, on the other hand, frequently necessitates such a demanding and stressful manoeuvre. This paper presents a training solution involving the use of a 360°full-mission bridge simulator.

1 SUPPORTING A 3D MULTI-STREAMER SEISMIC SURVEY OPERATION

1.1 *Seismic Vessel Capabilities*

Modern seismic survey operations are a far cry from the days of relatively small modified vessels towing a lone streamer or two that could be set and recovered in a few hours. Today, the seismic fleet is dominated by larger purpose-built vessels, though there are still many vessels in service, converted from other roles and designations. Modern seismic arrays typically comprise 12 streamers, possibly extending up to 8000 meters astern of the mother vessel, and measuring a swept path of 1200 meters across the ship's track. The very latest vessels now leaving the building yards have towing points for up to twenty such streamers. Such extensive equipment is capable of yielding a 3D seismic picture of great fidelity.

1.2 *The Seismic Survey Concept*

Seismic surveys are carried out extensively in ocean and offshore areas with a known potential for reserves of oil and gas in the sub-sea rock formations. The seismic survey vessel tows the steamer array suspended below the surface, carrying hydrophones. Sound waves are transmitted from the vessel using compressed air guns which travel down through the seabed and reflect back from the different layers of rock (Figure 1). These reflected sound waves are received by the hydrophones located along the seismic streamers which, when processed, gives a three dimensional picture of the substrata.

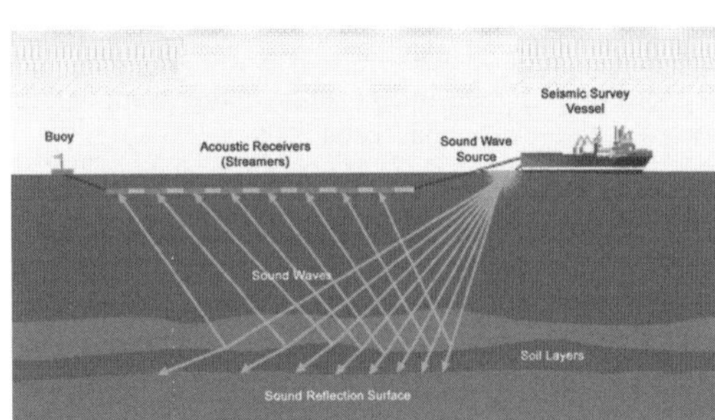

Fig. 1. Marine Seismic Survey

Seismic survey vessels 'in production' show the shapes and lights for a vessel restricted in its ability to manoeuvre. The streamer cables are spread by diverters/paravanes, similar in function to that of midwater trawl doors, and can extend to 1200 metres in width. The end of each streamer is marked by a tail-buoy carrying radar reflector and flashing lights.

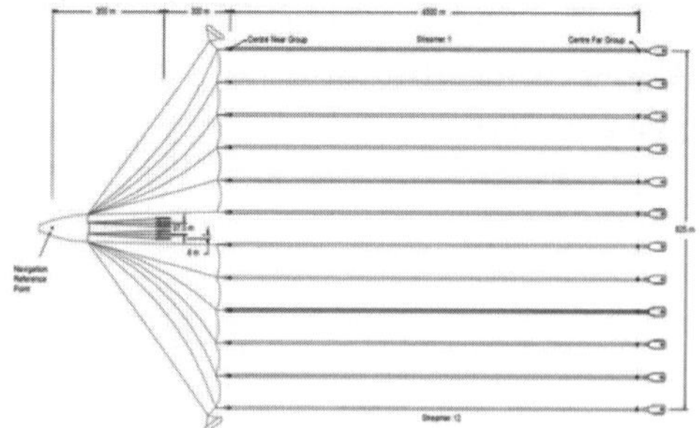

Fig. 2. A 12-streamer 3D Seismic Array

Seismic survey vessels tow at a speed of 4 to 5 knots and need to be on a straight line whilst surveying and are almost invariably accompanied by a 'Chase Boat' to police the immediate task vicinity and to assist in notifying other vessels of the seismic operation. Survey areas vary greatly in size and may cover extensive areas of the sea surface.

1.3 *The need for a support vessel*

It will be readily appreciated that the 3D multi-streamer seismic array described above is not something that could be deployed or recovered in a few hours: it is typically a two-day task, and perhaps longer. This means that once the survey operation has started and reached full production, significant interuption or suspension of the work is unlikely except for very pressing circumstances. A further consequence of that imperative is that the seismic ship must be supported and replenished under way, as the operational situation demands. An offshore supply vessel (OSV), suitably modified for the purpose, is the solution of choice for most seismic operators. And when not directly engaged on support and replenishment duties, the same OSV serves in the 'chase boat' capacity.

1.4 *The support vessel 'close alongside'*

The support vessel must replenish the seismic 'mother' ship with fuel and machinery consumables, primarily, but also with victuals and catering stores, and all other general and technical stores necessary for uninterupted seismic production. And from time to time the support vessel will be required to ferry and transfer personnel to and from the mother vessel, as when crew rotation is scheduled. These transfer operations (fuel, stores, personnel) are most often effected with the support vessel 'close alongside' the mother ship.

1.5 *The support vessel 'in-line'*

In other circumstances, and for various reasons, the support vessel may not be able to transfer fuel from a 'close alongside' position, in which case she will have to take station 'in-line' ahead of the mother ship. This demanding manoeuvre requires the support vessel to make a close approach 'in-line' ahead and bringing her transom to about 40 metres from the mother ship's bow, passing a shot line and messenger, then opening the distance between the two ships to 90 metres — a separation maintained by a heavy distance line — before passing the fuel line. The support vessel will now have the challenging task of maintaining her station for perhaps the coming six hours, until refueling is completed.

1.6 *Readiness to offer an emergency tow*

The support vessel will have one other major assignment in executing her close support role: to provide an emergency towing capability to the seismic vessel in the event of her suffering a serious propulsion failure. The consequences of such a power failure on the mother ship mean a very rapid loss of speed, caused by the drag of the paravane/streamer array. However, the inertia of this same array will ensure that its loss of speed is not as rapid as that of the mother ship.

1.7 *Short time window*

If forward motion is not restored to the mother ship she is in danger of becoming entangled and ensnared in her own gear – an expensive 'seismic spaghetti' of streamers, buoys, paravanes and towing lines. In such an emergency, the support vessel has a time window of perhaps 15 minutes (maximum) in which to make a close approach and connect up the emergency towline.

1.8 *Immediate priority*

The immediate priority for the support vessel is to get the stricken mother ship moving again, along the original path and away from the entanglement danger of her gear. Even one knot, or less, will achieve this and buy the time necessary to restore power on the seismic vessel.

2 SIMULATOR REQUIREMENTS

2.1 *The needs of the industry*

The initial request to develop a suitable simulation training programme for such unusual and unique replenishment-at-sea (RAS) manoeuvres came from a local offshore operator, Mainport Group, Cork. In late 2005, they won a contract from the French seismic operator, CGG (now CGGVeritas), to provide a seismic support vessel for operations in the Indian Ocean — and they needed simulator-based training for their crews. For the project, the 360° main bridge (full-mission) simulator at the National Maritime College of Ireland (NMCI) was utilized. Subsequently, Mainport added four similar vessels (OSVs) to their seismic support division, working for CGGVeritas and other seismic operators. Then in late 2009, CGGVeritas agreed with NMCI, a similar bespoke training programme for their seismic ship masters and mates, paired with their matching support vessel counterparts.

2.2 *Kongsberg full-mission simulators at NMCI*

The Kongsberg full-mission bridge simulators at NMCI are specifically designed for complex ship-handling manoeuvres and advanced navigation exercises. All important navigation and manoeuvring data are presented to the conning officer on the bridge via a comprehensive array of statutory (SOLAS) instruments and displays.

By configuring the 360°simulator with an accurately compiled ship model having realistic hydrodynamic characteristics, the high-end simulator generates a ship-manoeuvring environment of impressive fidelity. In this respect, the NMCI full-mission simulators are rated world-class, and the entire simulation centre is ranked amongst the best such facilities anywhere.

2.3 *Realistic 'ownship' behaviour*

In the simulated environment, the behaviour and response of the visual 'ownship' model to the engine, rudder and interaction forces and to the environmental conditions, is governed by a matching mathematical ship manoeuvring model. The model must behave in such a way that the position, heading, velocity and swept path of the 'ownship' are always representative of real ship behaviour.

2.4 *The necessity for a 360°simulator*

The panoramic visuals of a 360° simulator are an essential feature of any training exercise attempting to simulate the fidelity of a replenishment-at-sea operation. This constraint is self-evident for towing and 'in-line' RAS evolutions, where the support vessel master needs to have an unrestricted view astern. But it is equally valid in the 'close aboard' approach (support vessel approaching the mother ship beam-on) where the full broadside view is just as necessary.

2.5 *'Ownship' and 'target' models*

The 'ownship' model used as the simulated support vessel was the Kongsberg supply vessel SUPLY02L, fully dynamic in six degrees of freedom (heave, sway, surge, roll, pitch and yaw), representing all horizontal and vertical motions of the ship. For all RAS exercises the seismic mother ship may be simulated by a generic 'target' model, though a dedicated seismic ship 'target' model (CGG ALIZE, Figure 3) has recently been developed.

Fig. 3. Seismic Survey Ship CGG ALIZE

3 RAS PREPARATIONS

3.1 *Matrix of permitted operations*

A RAS evolution 'close alongside' the seismic ship requires planning, preparation, careful attention to procedure and skilful ship-handling. It is a daylight manoeuvre only, constrained by agreed limits on visibility, wind speed and direction, and sea/swell state — a maximum wind speed of 20 knots and significant wave height of 1.5 metres are the usual upper limits of acceptable conditions. Weather limitations and other restrictions for RAS and the wide variety of seismic operations should be promulgated in company manuals — CGGVeritas meets the requirement by publishing a tabulated *Matrix of Permitted Operations*, abbreviated to *MOPO* in their fleet guidelines and procedures. Tool-box meetings on both vessels are essential. Day, time and scope for the RAS transfer having been mutually agreed, either master must have complete discretion to abort the operation at any time during the approach phase or throughout the transfer.

3.2 *RAS speed of 4.5 knots*

Before any closing approach manoeuvre is initiated the mother ship must confirm her track, heading and speed, and throughout the RAS approach she must advise the support vessel of any small changes in those parameters — it is a given that any substantial changes should not be contemplated. The necessity of maintaining the operational speed for the streamer array dictates a typical RAS speed of about 4.5 knots.

3.3 *Why the beam-to-beam approach?*

In the most common type of RAS operation, involving naval formations, the re-supplying fleet tanker becomes the designated formation 'guide' and seeks to maintain her heading within 1°of the signalled replenishment course. And whereas naval vessels positioning for their RAS station will approach the 'guide' from astern or fine on her quarter, such aspect is never an option for a seismic/support RAS; the wide towline controlling the diverter/paravane is angled 45° outwards from the quarters of the seismic ship, which constrains the support vessel's approach to a narrow sector on the mother ship's beam.

3.4 *Choice of steering control*

The support vessel is likely to assume a standby station about 1000 metres abeam of the seismic ship, on the agreed side. A closer standby station is acceptable, but not if that station is inside the paravane path. There is much debate in the industry on the choice of auto-pilot or manual steering for the support vessel's RAS approach, especially in the latter phase of the manoeuvre when bringing the vessel from the 'close aboard' station (about 30–50 metres abeam) to the 'close alongside' position.

3.5 *The case for auto-pilot control*

The reality of small crews means that a skilled helmsman is unlikely to be available, particularly at a time of peak demand when all available crew are needed on deck. Also, given the widespread lack of opportunity for manual steering in commercial shipping there is deep concern within the industry that the manual steering skills of seamen, generally, are inadequate. Neither is it acceptable that the master should manually steer his ship in the final approach; he already has sufficient demands on his judgement, watching his speed, avoiding the wide tow wires and other overhangs and obstructions on the mother ship, controlling inter-ship and intra-ship communications, and, most critically, the ever-constant eye for interaction effects. In the circumstances, the case for using the auto-pilot is compelling, and no less compelling is the need to ensure that such equipment is fully serviced and totally reliable.

4 APPROACH AND DISENGEGEMENT

4.1 *Safe convergence*

In commencing her approach from the stand-by station, the support vessel must steer an inward convergent course (towards the mother ship) by about 20°, and increase speed so as to avoid increasing the aspect angle with the attendant risk of fouling the paravane tow wire. For instance, if the base course and speed (survey track) is 120° x 4.5 kn, a starboard-side approach will require the support vessel to steer 100° and set her speed at 4.8 kn. If the convergent angle were 30°, an approach speed of 5.2 kn would be necessary to maintain the same 90° aspect. A useful visual guide for the support vessel master is to keep the seismic ship's bridge-front in view: if, during the approach, an increased aspect angle leads to the loss of that view the support vessel has fallen abaft the optimum approach line, and runs the risk of fouling the wide tow wire.

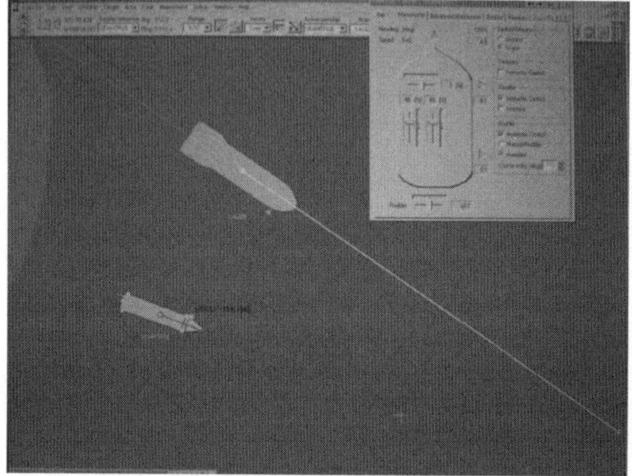

Fig. 4. Support vessel converging on mother ship

4.2 *Reducing the convergent angle*

As the support vessel closes the mother ship the convergent angle must be reduced (Figure 4). When the lateral distance between the ships is down to 100 metres the convergent angle should not exceed 10°, and at the same time the support vessel will need to trim back her speed; as the courses approach coincidence so too should the speeds of both ships.

4.3 *Suspend and reappraise*

The 100 metre mark is a good position at which the support vessel should temporarily suspend the convergent manoeuvre. This will allow the masters of both vessel the opportunity to reappraise the situation and to reassure each other that all checklist parameters for a safe RAS operation remain valid.

Fig. 5. Both vessels confirm willingness to proceed

4.4 *Coming into position 'close alongside'*

If there is no reason to abort the evolution the support vessel should resume the convergent course (Figure 5). Ten degrees convergence is still acceptable, but this should be gradually reduced so that as the support vessel arrives in the 'close aboard' station, 50 metres from the mother vessel, the convergence should not exceed 5°. In the final phase of the manoeuvre, from 'close aboard' to 'close alongside' the convergent angle must be reduced further so that when contact is made on the yokohama-fenders the convergence is 2° or less. In reality, when the ships are 10–15 metres apart there is usually little need for any convergent angle because the dominant interactive force between the ships at this stage is most likely to be that of attraction.

Fig. 6. 'Close alongside' replenishment station

4.5 *Avoiding simulation 'freeze'*

In the normal course of simulation exercises the intuitive simulator response to the 'ownship' model making contact with a 'target' vessel at close quarters is to signal a collision condition, at which point the exercise functions freeze and the simulator must be reset. In the RAS simulation the same outcome is evident when the support vessel (as the ownship) makes heavy contact with the mother ship, such as when the convergent angle is too large or when her position alongside is so far aft that she strikes the protruding sponson structure. On the other hand, if the contact force between the ships is gentle and correctly positioned, as when there is little or no convergent angle, the simulation exercise will continue without interruption (Figure 6).

4.6 *Critical securing lines only*

Once the support vessel, properly fendered, is safely alongside the mother ship the agreed securing/mooring lines must be rigged. These usually consist of a for'd breast-line and two fore-springs – lines connected from the forepart of the support vessel only. Bearing in mind the formation speed of 4-5 knots, it is never acceptable to have any securing lines connected to the after-part of the vessel.

4.7 *Favourable disengagement*

When fuel and stores transfer is completed the support vessel must prepare to disengage and clear away from the seismic vessel — all before the onset of darkness. On some seismic/support vessel combinations the vessels will separate and diverge under the favourable effect of hydro-dynamic interaction, as soon as the securing lines are released. But in many cases, this will not happen.

4.8 *Adverse interaction effects*

Where the adverse interaction effects are dominant, the force of attraction between the ships will restrain the support vessel in the 'close alongside' position. Any attempt by the support vessel to clear the side of the mother ship by increasing speed and using outward helm will fail because it is not possible to steer away in these circumstances. If the attempted manoeuvre is allowed to continue, the outcome illustrated in Figures 7 and 8 is inevitable; the support vessel will move forward, all the while restrained against the side of the seismic ship, until the critically adverse interaction effect becomes manifest. This is the 'bow-in' turning moment that will cause the support vessel, despite carrying outward helm, to turn across the bow of the mother ship. Averting disaster at this point is in the hands of the seismic master, who must take all way off the vessel instantly.

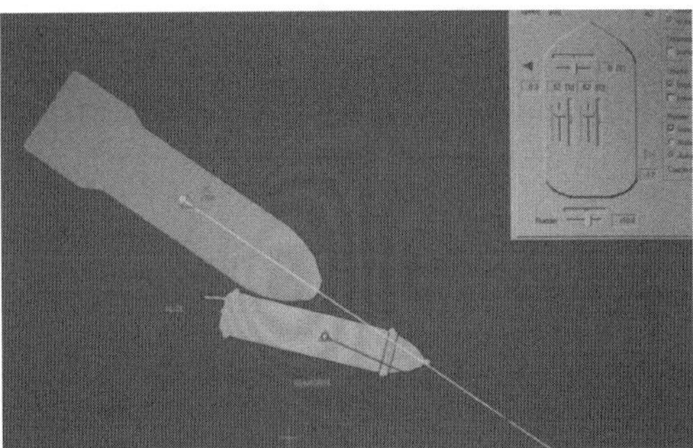

Fig. 7. Critical interaction effect

Fig. 8. Imminent risk of capsize

4.9 *Using thruster and inboard helm to overcome adverse interaction effects*

Where, as described above, interaction effects prevent the support vessel from steering directly on a divergent course from the 'close alongside' position she must first use her bow thruster to open out a divergent angle. As this angle increases the master will need to increase speed and apply 5° inboard helm (to keep the transom clear of the side of the seismic vessel). How long to keep the inboard helm applied is a judgment call, but it should not be maintained if its turning force threatens to overcome the outward rate-of-turn from the bow thruster. Once the support vessel is 10–15 metres clear of the mother ship and has a divergent heading of about 10° she has the manoeuvring freedom to steam clear away with little risk of exposure to any further adverse interaction effect.

5 TOWING

5.1 *'In-line' RAS station*

In the circumstances where the support vessel is unable to transfer fuel from the 'close alongside' position, she will have to take station 'in-line' ahead of the seismic ship. This manoeuvre requires the support vessel to make a similar approach to the 'close aboard' station, as described above, and then increase speed while holding a convergent course. The objective is to pass within 50 metres of the mother ship so as to take temporary station 'in-line' ahead of her, with about 40 metres separation between her transom and the mother ship's bow. This close proximity facilitates the exchange of a shot line, messenger and 'distance line'. The distance line is of hawser-like quality and is used to maintain a near-constant distance of about 90 metres between the two ships. Once the heavy distance line is established and lightly tensioned, the fuel transfer line is then rigged between the ships. The support vessel must now settle into the stressful role of constant vigilance in seeking to maintain her station for perhaps the coming six hours, until refueling is completed.

5.2 *Towing exercise requires second simulator*

The simulation of an emergency tow scenario requires a different arrangement of simulators and models. While a tow-line may be assigned and controlled from the support vessel 'ownship' it is only possible to connect it to another 'ownship', which, in turn, must be assigned to another simulator. A further problem arises in the simulation exercise when attempting to achieve towing fidelity. An actual support vessel confronted with an emergency tow scenario will need to use substantial power to get the seismic vessel moving at just 3 knots, because of the enormous drag created by the streamer array and associated gear (approximately 80t). A simulator 'ownship' assigned as the seismic ship will generate only the drag appropriate to the particular model dynamics. However, if the simulator ownship control includes the optional External Forces menu it is possible to apply a range of such forces to the seismic ship model so as to achieve realistic fidelity in simulating seismic streamer drag.

6 CONCLUSIONS

The modern multi-streamer 3D seismic survey operation is enormously challenging, in the financial and technical resources required to mount and maintain the venture at sea. Downtime in seismic production carries significant penalties, hence the need for the unique OSV support described in this paper — support activity for which few mariners are likely to have prior knowledge or experience. A properly resourced full-mission 360° simulator centre is able to meet that specific training need.

REFERENCES

International Association of Geophysical Contractors (IAGC), 2002, *Marine Seismic Operations: An Overview.*

Naval Warfare Publication, 2004, *Underway Replenishment NWP 4-01.4,* US Navy Department. Available at: http://www.navybmr.com/NWP%204-014.html

Maritime and Coastguard Agency (MCA), 2002, Marine Guidance Notice *MGN 199 (M) Dangers of Interaction*, Southampton, MCA.

Paffett, J. 1990, *Ships and Water.* London, The Nautical Institute.

McTaggart, K. Cumming, D. Hsiung, C. & Li, L. 2001, Hydrodynamic Interactions Between Ships During Underway Replenishment, 6th Canadian Marine Hydrodynamics and Structures Conference, Vancouver, 23-26 May.

Skejic, R. Breivik, M. Fossen, T. & Faltinsen, O. 2009, *Modeling and Control of Underway Replenishment Operations in Calm Water*, 8th IFAC International Conference on Manoeuvring and Control of Marine Craft, Guarujá (SP), Brazil, 16-18 September.

18. Teaching or Learning of ROR

V. K. Mohindra & I. V. Solanki
Applied Research Institute, Newl Delhi, India

ABSTRACT: It is a well established fact that many of the collisions at sea were the result of either lack of proper knowledge of rules to prevent collision at sea or their improper application. Safe navigation and collision avoidance are the prime responsibilities of an officer of watch. The OOW must therefore be competent to the extent of zero tolerance. But why is it not so? Two teachers try to look at the present method of providing and testing the level of competency and to suggest changes for ensuring that ROR is given a greater emphasis than at present. After all the teacher is last link of interface with students and is the best source of feed back on why a student would not want to or fail to learn ROR. A student's success is a teacher's success.

1 BACKGROUND

1 A study by underwriters put percentage of collision/collision losses as under:

Period	Total Losses	Contact/Collision Losses
92-96	620	63 (10%)
1996	179	29 (16%)

2 MAIB annual report for 2004 reported 33 collisions of which almost 80% were due to lack of competency - a very alarming situation indeed.

3 More studies must have since been carried out with more accurate data. But the fact remains that, considering the size of vessels involved and likely damage to property and environment by maritime accidents, even one collision is one too many.

2 CAUSES OF ACCIDENTS

1 In 1976 the US National Academy of Science had identified 14 major causes of human error in Maritime Accidents as under (not in order of priority):

- inattention
- ambiguous master-pilot relationship
- inefficient bridge design
- poor operational procedures
- poor physical fitness
- poor eye sight
- excessive fatigue
- excessive alcohol use
- excessive personal turnover
- high level of calculated risk
- inadequate lights and markers
- misuse of radar
- uncertain use of light signals
- inadequacies of ROR.

2 Bridge design, working conditions and compulsory rest hours, have since been addressed though some still take short cuts. Master-Pilot relationship is being taught every where and more attention is being given to physical well being of the crew. Large crew turnover is the result of outsourced manning system and can only be tackled through joint efforts of ship owners, operators, and underwriters.

3 The last five factors can be grouped under rules to prevent collision at sea or ROR, as is commonly called.

3 WHAT HAVE WE DONE ABOUT IT

1 Inadequacies of ROR have seen a very animated debate with Capt Roger J Syms of Australian Maritime College spearheading the demand for a change. He conducted a number of surveys and tests which pointed out serious gaps in knowledge of ROR and its application amongst his students.

2 Grey areas in the current rules were discussed in detail during a seminar held in China. Leading authorities on ROR like Capt Ian Cockroft have added a rider that current rules should only be substituted with the new rules that have fewer or no grey areas and not just different grey areas. It is not our intention to discuss merits or demerits

of the call for change and we do wonder if we can ever have perfect rules.

3 Capt Roger J Syms also tested a class on knowledge of ROR. Half the class could refer to the text book while answering. A large number of examinees gave incorrect answers) gave wrong answers (over 80% in case of rule 19) including those using reference books. The methodology or accuracy of these statistics is not important. A few less here or more there do not really matter. However no one can question that all of the above data sends a loud and clear message that:
   - Collisions do form a major part of maritime losses
   - ROR does play a major role in human failure during collisions.

4 THE FOUR FACTORS

1 Four factors involved in learning of ROR are:
   - Rules
   - Syllabus
   - Examination
   - Teaching

2 Historically, codified ROR goes back to 5th century AD. It went through various stages and revisions and has been under the control of IMCO/IMO since 1960. The process will continue as and when necessary.
   But making rules is only one half of the story. Rules, as they exist today or are amended in future, with or without their flaws, have to be learnt by seafarers manning ships at sea in a manner that will ensure an instant and correct response authorised by the rules to any possible threat of collision or close quarter situation and ensure safety of both vessels along with other watch keeping duties. This is where the syllabus comes in because it will guide the teaching.

3 Method of examination sets teaching and learning goals because students only focus on what is needed to pass the examination and resist any extra work.

5 CURRENT SCENARIO

1 The old traditional teaching method required students to memorize rules and examiners often asked candidates to repeat rules in support of their answers. Post STCW regime has changed rules of the game.

2 STCW requires:
   "Thorough knowledge of content, application and intent of International Regulations for Preventing Collisions at Sea"

3 This has perhaps led to the belief amongst teachers and examiners that memorizing rules is not necessary. The syllabus, teaching and learning objectives are framed accordingly.

4 Semester system adopted by most Maritime Universities is another hindrance as it does not permit repetition of credits.

5 Random sampling system of examination for competency is not really a fail proof method.

6 WHAT IS NEEDED?

The multitasking officer on watch (OOW) needs such level knowledge of rules that it becomes a second nature. He has to be capable of instant situation analysis and collision avoidance action. To achieve this, a student must breathe, eat and dream ROR.
This is the challenge for framers of syllabus, examiners and most of all for teachers.

7 TO MEMORIZE OR NOT

1 During fifties, students were expected to memorize the rules verbatim and know their application to situations. They were tested across the examiner's desk using small wooden models. The candidate had to explain the situation, decide on action he would take and justify his action by quoting rule number authorizing the action and sometimes even repeating it verbatim. Some examiners even asked the expected action of the other vessel.
   I memorized the rules from almost day 1, and never ever had the experience of jelly beans in my stomach whilst on navigation watch except once when an intentional action almost went wrong. The rules are engraved in my memory.

2 Unfortunately, it does not seem to work any more. Students resist memorizing the rules as they are told from day 1 that they only need a thorough understanding. But how does one get this thorough understanding?

3 Before arrival of GPS, we memorized formulae for calculating position lines. There was no question of just a thorough understanding. Can we understand the Pythagoras theorem without memorizing that in a right angle triangle the sum of squares of the base and the perpendicular is equal to the square of the hypotenuse?

8 WHY SHOULD WE NOT MEMORIZE?

1 Some argue that memorizing rules takes up a lot of space of our memory. But so do names, telephone numbers, addresses etc. What all should we erase or avoid memorizing? A number of facts like an accident we witness, news headlines or a beautiful smile etc. get stored in our memory even without a deliberate attempt.

2 Memories of computers are measured in terms of their smallest addressable element, called a byte. A byte usually contains eight binary digits. Nerve cells also have an "all or nothing" binary response. If combinatorial codes are remembered by nerve cells, each combination of firing inputs received by a neuron with 100 dendrites could contain 100 binary digits. The possible number of unique combinations of inputs for a single neuron with just 100 incoming dendrites could be computed as 100 x 99 x 98 x 97 x 2 x 1 .possibilities. That represents more than 1, 000, 000, 000, 000, 000, 000, 000, 000, 000, 000, 000, 000, 000, 000, 000, 000, 000, 000, 000, 000, 000, 000, 000, 000, 000, 000, 000, 000, 000, 000, 000, 000, 000, 000, 000, 000, 000, 000, 000, 000, 000, 000, 000, 000, 000, 000, 000, 000, 000, 000, 000, 000, 000, 000 unique possible combinations! Multiply that number by 100 and divide by 8 to measure the number of bytes of possible memory. A single nerve cell with 100 dendrites can potentially remember that many bytes of singular combinations. Some nerve cells have up to 2,50,000 dendrites! Only the possible existence of such codes can explain the phenomenal capacity of human memory.

3 Robert Birge (Syracuse University), who studies the storage of data in proteins, estimated in *1996* that the memory capacity of the brain was between one and ten terabytes, with a most likely value of 3 terabytes. 1 terabyte is 1000000000000 bytes, or 1 trillion (short scale) bytes, or 1000 gigabytes).

4 And entire ROR only takes up approximately 500 kB – 1 mB of space?

5 Medical profession requires far greater amount of memorizing but that does not affect proficiency of doctors or surgeons. Study of History or Languages, is no different.

6 Yet another argument is that it is the application that is important. Yes but we must first know what to apply. A vessel crossing on our starboard bow involves rules 7, 8, 9, 10, 15, 17 and 18. They must be in our memory bank for instant access, analysis and conclusion.

7 Finally, language difficulties are sometimes quoted as an impediment to memorizing. I did manage to get students in Iran to memorize the rules that I considered important.

9 A TEST

1 Let me now recount my experiment with one group of cadets. They were given 30 minutes to learn rule 1(a) and to then write it in their own words. Some reproduced it, some produced longer versions of the rule and some missed out important parts like `navigable by seagoing vessels´. In short there were so many different versions amongst a group of only 40 English proficient cadets. I tremble at the very thought of what to expect amongst a larger group including non English proficient candidates.

2 In rule 9(a) words `proceeding along the course of a narrow channel or fairway´, if ignored, change the entire meaning of the rule and can lead to chaos. Same applies to words `power driven vessels´ and `so as to involve risk of collision´ in rules 14 and 15.

10 CONCLUSION

1 In conclusion, I believe that examiners and teachers bear the final responsibility to ensure that only persons with absolutely thorough knowledge of rules and their application serve onboard ships. Our objective must be to ensure that students learn rules and not to just teach rules.

2 Memorizing will ensure the thorough knowledge and extensive exercises on proper simulators permitting visual bearings will demonstrate application ability. These must be repeated and repeated so as to leave an indelible imprint on the students´ minds. Nothing short of it should be acceptable.

3 It is not enough to look for teaching methods for achieving this zero level of tolerance. We have to create an environment so that students will want to do so.

11 RECOMMENDATIONS

1 Some recommendations:
- Identify rules that must be memorized to the extent of total verbatim recall.
- Arrange syllabus so as to repeat rules in every semester.
- Devise methods to test total recall as well as understanding of rules.
- Use simulators extensively to teach and test application of rules.
- Adopt zero tolerance towards proficiency in ROR while teaching and examining.

2 Many amongst us will differ strongly and perhaps for very valid reasons. Let it be debated.

3 This is the main objective of this presentation – Teaching/Learning of ROR and we hope that this presentation will set us on a course of a global discussion amongst all teachers and lead to a common most effective method of ensuring acceptable level of knowledge amongst all seafarers.

4 Let us, the teachers, start a movement aimed at improving teaching and learning of ROR which will dictate changes upstream. Let the teachers set the Agenda for examiners, designers of syllabus

and finally STCW. Let is change the environment from teaching to wanting to learn so that the student demands to be taught.

12 BOTTOM LINE

No one, who can be a potential threat to shipping, should be permitted onboard a ship. It is a tall order but not impossible. We can do it.

REFERENCES:

1. Rules for 21st Century, Roger J Syms, Seaways September/October 1994
2. Research Needs To Reduce Maritime Collisions, Rammings and Groundings, 1981, National Research Council Washington D C Maritime Transportation Research Board. (Maritime Transport Research Board Commission on Sociotechnical Systems)

19. Safety and Security Trainer SST_7 –A New Way to Prepare Crews Managing Emergency Situations

C. Bornhorst
Maritime Account Simulation Products, Rheinmetall Defence Electronics GmbH, Germany

ABSTRACT: Emergency Response, Crew Resource and Crisis Management are one of the most important parts in education and training of nautical officers and engineers. Increasing size of cargo and number of passenger ships, decreasing number of crew members, highly complex electronic support systems and additional threats in some dangerous waters necessitates better preparation for unforeseen maritime emergency cases at all times. This can also be seen in the new Manila Amendments coming into force on 1 January 2012. But how can crews be prepared and trained accordingly, without complex and costly training scenarios, because especially the right behaviour and communication under stress situations in regard to maritime emergency cases like fire on board, water inrush, evacuation or maritime piracy is difficult to be trained.
It is already accepted worldwide, that modern simulation systems can support training more effectively, intensively and economically. Therefore recently Rheinmetall Defence Electronics introduced a totally new type of simulators which is perfectly matched to these requirements – the new "**Safety and Security Trainer SST_7**" This simulator is focused on management and decision of emergency scenarios and provides exercises which have to be managed by an emergency team under conditions close to reality. The system is based on 3D serious gaming technology and provides a high degree of realism. The main goal of this simulator is to train how to analyse, decide, conclude and communicate in emergency situations.

1 INTRODUCTION

For some years, besides the typical threats of fire and water inrush, the greatest risks at sea are maritime terrorism and piracy. Therefore safety and security issues are thus a high priority, as also shown in the new STCW Manila Amendments with its considerably tightened safety and security requirements signed recently.

Among others this results in a growing demand for training and education in this field. Universities, academies and other education centres as well as companies operating in maritime environment have to prepare themselves to mange all kinds of maritime emergencies. Especially an increasing demand on emergency training providers is recognisable, to ensure that, in the event of an emergency situation, the onboard management level is capable to make the right decision just in time. In this context, damage limitation can only be realised, if competent crisis management is enacted by the crew. But how can the crews prepare themselves for these emergency situations that will hopefully never occur? How can the cooperation between nautical and the technical crew be practised? In the course of conventional practical safety training, the focus is more on practical handling of safety equipment, whereas until now team training at management level has only been possible in classrooms. At that point the new generation of simulator provides a significant improvement. In a crisis situation, the following skills are required from the leadership on board:

- fast competent action,
- correct appraisal of the situation,
- effective communications structures,
- a strategy to maintain operation of the ship,
- protection of persons and the environment.

The most important aspect in managing a crisis is a fast, communicative and professional collaboration of the crew members. Each crew member must know and fulfil his function in accordance with the safety role. However, this can only function, if the command structures, safety and security processes and communications are clear and established.

In a crisis situation, each crew member must immediately remember safety rules and processes and has to act accordingly. This can only be achieved again and again if preservative safety exercises are carried out frequently and as close as possible to reality. But in times of economic pressure and rationalisation, these safety exercises cannot be carried out as frequently as necessary. Furthermore, in many safety exercises the situation is not close enough to reality, crew members are not taking it seriously and stress is missing which makes it more difficult to act correctly in a crisis situation.

In order to provide a solution by means of a high fidelity and professional simulation system Rheinmetall Defence Electronics GmbH / Germany started the design and development of its.

Maritime Safety and Security Trainer SST_7

Figure 1. Example for a room layout, safety office including beamer in centre

Today RDE is able to offer the first holistic training and education system for Safety and Security at Sea. The concept covers nearly all aspects of Safety and Security like fire on board, inrush of water, collision, running aground, Person Over Board, evacuation, damage/destruction of ship, hijacking or seizure, attacks, use of ship to cause incidents or as a weapon etc.

The system provides a perfect combination of simulated emergency scenarios including introduction of technical malfunctions and a most realistic training environment.

The new **Safety and Security Trainer SST_7** comprises amongst others the following highlights:

- covers all ISPS required training aspects, from Advanced Fire Fighting to Ship Security / Company Security Officers Training and Crowd and Crisis Management
- is in accordance with new Manila Amendments
- Safety and Security Trainer can be configured in two different modes: team training and individual student training
- modern 3D visualisation of training platforms based on serious game engine, e. g. container vessel or Ro/Pax ferry
- own ship model can be tailored to customers requirements
- for greater lasting effects of training, debriefing with replay of exercises and assessment is possible

The **Safety and Security Trainer SST_7** can be configured for team training or individual student training. In team mode the students are situated on one and the same ship, so communication, interactions and decision making of an emergency situation can effectively be trained. Alternatively to a team, students can individually train their skills in their own exercises on their own ship.

Access to every deck by a detailed graphical 3D representation, complete functional systems (electrical systems, engines, ruder, etc), communication possibilities as aboard a real ship and everything in real time characterises are the major functionalities of the simulator.

2 SST_7 CONFIGURATION

Onboard of most ships the crew, involved in managing emergency situations is organised in a bridge team, ship safety office and local teams. This kind of organisation was the model for the **SST7** and is reflected in corresponding **SST7** installations: instructor and trainee workstations organised in bridge, safety office and local teams. Each station consists of two monitors, one being the situation monitor, the other being the action monitor, keyboard and mouse to guide a virtual person and a headset for communication.

Figure 2. Instructor / Trainee Station

The instructor, who is setting-up the simulator, pre- paring, starting and monitoring the exercise as well as conducting the debriefing afterwards, can start the system in different modes: team training or student training.

In team training the same assignment is used as onboard, one workstation for the bridge team, one for the ship safety office and several for the local teams. In addition the instructor can also participate in the exercise, e. g. as an additional crew member, passengers or as a shore based unit. In student training, each trainee has to act on his own and has to manage the task by himself. He is assigned to a ship and he has to play all roles. In both modes he can introduce malfunctions and controls the entire communication, including recording of data and communication.

3 EXERCISE PLATFORM

The **SST$_7$** is based on a virtual 3D ship model. Currently, two models are available, a 4500 TEU container carrier and a combined RoRo / Passenger ferry. Both ships are fully modelled (all desks) and type free, which means, the ships are in accordance to real ships, but not representing a specific one (if requested, specific ships´ can be modelled).

Each trainee, assigned to a workstation, is representing one crew member. The crew member is guiding a virtual person called avatar. By using the keyboard and the mouse, the trainee is able to direct his avatar through the ship, executing operations, like analysing sensors or using fire fighting components etc. During the whole exercise, every crew member is able to communicate accordingly, either direct, by intercom or walkie-talkie.

Figure 3. View on cardeck of a Ro / Pax Ferry

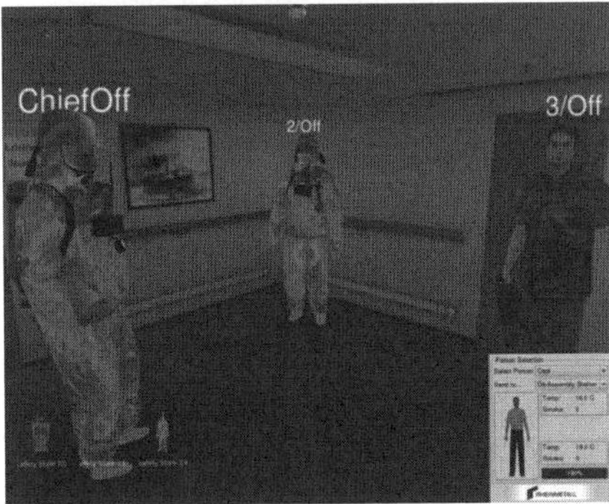

Figure 4. Three "crew members" during an exercise

4 COMMUNICATION

A sophisticated VoIP (Voice over Internet Protocol) Communication system allows communication as in reality. Crew members can talk to each other, if their avatars are located in the same room. If not, communication has to be done by intercom or walkie-talkies, if the avatar is equipped with one. Also public announcements can be done. The complete communication is recorded and can be used for detailed debriefing.

5 PHYSICAL MODELS

The backbone of the complete **SST$_7$** simulation are various mathematical/physical models. The highest goal of the development is, to simulate all processes as close to reality as possible, in order to provide the most realistic training environment. Main physical models are for:

- ship systems,
- automation systems,
- fire, incl. smoke,
- fire extinguishing,
- sensors,
- water inrush, incl. stability,
- human health,
- safety equipment.

Derived from RDE´s well known and mature Ship Engine Simulator **SES$_7$** important ship systems (e. g. propulsion, ballast system, power supply) are implemented, completed by various automation systems.

The fire model is based on thermodynamic calculations allowing a fire simulation close to reality. Oxygen and CO_2 are taken into account as well as propagation, thermo conductivity and other effects. For example this results e. g. in effects like flash over and back draft. Fire on board can be extinguished with different fire fighting material. Water, foam and powder are at the trainee's disposal as well as CO_2 and FM 200. A60 walls and doors are modelled and implemented. The complete ship model is divided into fire cells. Each fire cell can be programmed by the instructor, in relation to the kind of material, amount of material, start of fire etc.

Cracks or holes can be defined by diameter or length. Water inrush has influence on stability, the ship can capsize or even sink. Safety measures like water tight compartments, bilge pumps or the ballast system can be used and operated. The health status of the avatar is continuously indicated and if a trainee is not taking care, the avatar can be seriously injured or be killed.

Typical safety equipment is modelled, up to a water drenching system (Ro/Pax). A lot of protection and counter activities can be selected like fire hoses and extinguisher, CO_2 system survival suit and even gas detectors.

6 SAFETY AND SECURITY PROCEDURES

Apart from already mentioned safety processes like fire fighting and water inrush other accompanying procedures like evacuation and security related processes can be trained. To support these functions ship security alarm and a ship security room is modelled. The crew can train how to behave in case of a security threats. In addition protection suits, explosimeter and gas detectors can be used in order to train search and rescue routines in order to identify explosive or toxic materials.

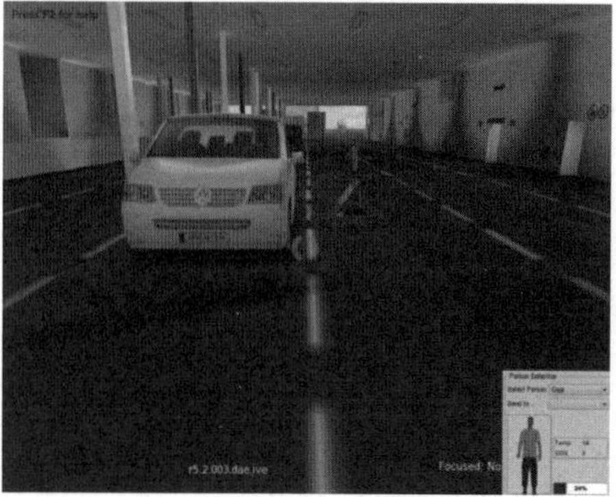

Figure 5. Mini robot "Telemax" investigates a car

7 INTERFACE AND SUPPORT SYSTEM

To make the **SST_7** even more valuable, the system can be interfaced to other simulation systems. A Ship Handling Simulator can be connected as well as a Ship Engine Simulator. This allows even more complex team training scenarios e. g. for Emergency Response Training.

Furthermore **SST_7** can be linked to an original Decision Support Systems, e. g. MADRAS, a product produced by the German company MARSIG. Together with these components, high level emergency team training can be conducted providing the highest value for training and the crews.

8 REGULATIONS AND CERTIFICATION

SST_7 was designed in accordance to STCW 95 and takes into account about the recent Manila Amendments, ISM/ISPS and SOLAS regulations. The system is certified by Germanischer Lloyd and DNV and was achieved by support of University of Applied Science Wismar / Warnemuende and company MARSIG mbH, both located in Germany.

9 CONCLUSION

Until today there is no comparable system on the market, providing maritime Safety and Security training for management levels. **SST_7** provides an unique and holistic concept for emergency training and assures all users a high standard of qualification without complicate, complex, time-consuming and costly simulation set-ups. **SST_7** opens a new door for modern and efficient training, provides professional trained crew members, protecting human life and valuable cargo as well as the environment against injuries and damages.

Rheinmetall Defence Electronics (RDE) is an internationally recognised supplier for the development of state-of-the-art systems and products for maritime applications and efficient simulation and training systems. The simulation systems cover the entire range of simulation systems - from Computer Based Training (CBT) to full mission simulators, including part task trainers, tactical simulators and appropriate courseware.

Within the business division Simulation and Training RDE offers a wide range of simulators for merchant marine and naval applications, including navigation and radar simulators, marine engineering and heavy lift and offshore simulators. Experienced in manufacturing and installation of Integrated Maritime Training Centres RDE's simulators provide students, trainees and seafarers in Europe and around the globe with an excellent environment for initial and advanced training. RDEs simulation systems exceed customer's demands, requirements and official regulations.

20. MarEng Plus Project and the New Applications

B. Katarzyńska
Gdynia Maritime University, Gdynia, Poland

ABSTRACT: In the MarEngPlus project the partner organisations from several different European countries produce new sections into an earlier developed web based Maritime English Learning Tool MarEng. MarEng Plus project started in 2008 and finished in 2010 and is complementary to the MarEng as it provides the materials at an elementary level and includes two new topics: Maritime Security and Marine Environment. As a result of the project the MarEng learning tool was transferred to new user groups and geographical areas. The new material not only widens the overall user group, but also motivates lower level learners to learn maritime English. In addition, the new partners in the project gained project work knowledge and experienced the process of creating a language learning tool in a cooperation project.

1 DEVELOPING THE MARITIME ENGLISH LEARNING TOOL

MarEng Plus project is a continuation of the MarEng project which was completed in 2007 and has been successfully used as a Maritime English teaching and learning tool in nautical schools and maritime universities since that time. MarEng Plus project started in 2008 and finished in 2010 and is complementary to the MarEng as it provides the materials at an elementary level and includes two new topics: Maritime Security and Marine Environment.

The two new topics in general and Maritime Security problems in particular, are very much in the news due to the frequent piracy attacks on cargo vessels, and that is the reason for the section in theMarEng Plus project which deals with the issues of Stowaways, (at an elementary level), the International Ship and Port facility Security code (at an intermediate level) and Ship Security Assessment (at an advanced level). All these topics have been accompanied by a number of different exercises which have all been recorded.

2 THE PROJECT PARTNERS

The partner group consists of a wide variety of maritime institutions, and involved in the project are education and maritime experts such as English teachers, researchers, training managers, seafaring professionals and representatives of the maritime industry. MarEng Plus is partially financed by the Leonardo daVinci programme of the European Union. The partners to the project come from seven different countries and include:

- University of Turku, Centre for Maritime Studies in Finland,
- Kymenlaakso University of Applied Sciences,
- Aland University of Applied Sciences
- Lingonet Ltd (software applications)
- Antwerp Maritime Academy, University of Antwerp, Belgium
- Cork Institute of Technology – National Maritime College of Ireland,
- Latvian Maritime Academy in Riga, Latvia
- Gdynia Maritime University, Poland
- University of La Laguna, Spain
- Shipping and Transport College, Rotterdam, Holland

The Mareng Plus project materials have all been tested and evaluated by a number of advisory partners which include:

- Finnish Maritime Administration,
- Finnish Port Operators Association,
- Emergency Services College in Finland,
- Estonian Maritime Academy,
- Latvian Maritime Administration,
- Lithuanian Maritime Academy,
- Baltic Ports Organisation,
- Ceronav Maritime Training Centre, Romania
- Maritime and Fishing School in Spain,
- Dokuz Eylul University in Turkey,
- IFAPA Centre in Huelva, Spain,
- IPFP Maritime & Fishery School, Canary Islands, Spain,

- Turkish Maritime Education Foundation, Institute of Maritime Studies, Turkey,

The external evaluator of the MarEng Plus project is prof. Clive Cole from the World Maritime University in Malmoe, Sweden.

3 WHAT'S NEW IN MARENG PLUS

The existing MarEng Learning Tool consists of intermediate and advanced level learning material on different maritime topics. The material is based on an idea of a virtual vessel that during its journey encounters different language usage situations in port and on board.

The aim of creating new material into the MarEng Learning Tool is to widen the user base of the Tool. As a result of the MarEng Plus project, two new topics, Transport security and The Environment, as well as elementary level learning material and a Teacher's manual, were added in the Tool.

4 NEW ELEMENTARY LEVEL

The MarEng Plus project offers a new elementary level which includes:

- Cargo Operations
- The Engine Room
- The Navigation Bridge
- Radio Communication
- The Weather
- First Aid
- Severe Weather Conditions
- The Marine Environment
- Maritime Security

The MarEng Plus project also includes two new topics at intermediate and advanced levels:

- Maritime Security
- The Marine Environment

It also includes Teacher's Manual and Answer Key. Another new feature is the mobile phone application of the Maritime English glossary.

The MarEng Plus project materials can be used online and are freely available on the internet at http://mareng.utu.fi.

5 MARENG WEBSITE

To enjoy the new MarEng Plus material, you should order a copy of the CD-ROM for free or download it on your computer from this site.

All of the MarEng Plus material is English-English and is based on language used in actual situations on board ships, in ports and elsewhere in the shipping chain. The new material will not only widen the overall user group but also motivate to learn maritime English in different levels.

Figure 1. MarEng Plus project logo

6 CREATING ELEMENTARY LEVEL AND TEACHER'S MANUAL

The feedback has also revealed that the lower level English learners are in the need of a (beginner) elementary's level as the MarEng learning tool currently consists of only intermediate and advanced levels. Teachers using the MarEng tool see that their teaching process could be made more efficient by creating a teacher's manual. Therefore, creating of elementary level and a teacher's manual will be a part of the project.

7 CONCLUSIONS

To conclude, the section on Maritime Security is particularly important and can be used as course material or complementary material in designing courses on Maritime Security which are vital in preparing the students and seamen for problems such as piracy, stowaways, ISPS code implementation and Ship Security Assessment which they may come across in their work on board vessels.

The MarEng Plus project fills in the gap and addresses these aspects of Maritime Security and both the students and the teachers are welcome to access the website at http://mareng.utu.fi and use the materials in practice. The MarEng Plus project was partly funded by the EU Leonardo da Vinci programme and that is why it is free of charge.

During the ppt presentation, time permitting, the author would like to show one of the sections on Maritime Security e.g. the section on stowaways and some of the exercises.

REFERENCES

http://mareng.utu.fi/ - official website of MarEng Project.

http://www.dnv.com/industry/maritime/servicessolutions/statutoryservices/isps/ - DnV ISPS Code website.

http://human-rights-convention.org/ - official website of European Convention on Human Rights.

http://www.un.org/en/rights/ - official website of the United Nations Declaration on Human Rights.

http://www.spc.noaa.gov/ - NOAA (National Oceanic and Atmospheric Administration) severe weather.

http://www.mapsofworld.com – Maps of World website.

21. Methods of Maritime-Related Word Stock Research in the Practical Work of a Maritime English Teacher

N. Demydenko
Kyiv State Maritime Academy, Ukraine

ABSTRACT: The article purposes to investigate the problem of Maritime English training to the 1st and 2nd year students of the Faculty of Navigation at a higher Maritime institution in connection with maritime – related word stock teaching process. The linguistic and methodological approaches to the problem make possible to open the discussion of the interdisciplinary strategy, i.e. the specific roles in practical work of specialists' department and that of ME department.

1 GENERAL NOTIONS

In today's MET education the domination of skills development over knowledge delivery is being obvious. This is the sphere of practical methodology which usually has great achievements in non-English speaking countries but depends a lot on national tradition existing in the system of education. Different methods, approaches and techniques used by practical specialist lecturers and language teachers sometimes result in insufficient language proficiency. Unfortunately, it happens in case of seafarers who have an opportunity of using ME for professional purposes in the process of studying thus comparing their language proficiency with that of other members in a multilingual crew. Here comes the point of coordination between professional and language teachers and a very sensitive matter such as methodological priorities of each of the parties. This is the core idea of any corporative research activities: Who does what? What are the domains of professional teachers and those of language teachers? In fact, there are so many questions which should get their answers, that the problem of up-to-date efficient practical ME methods and materials corresponding to the international standards for different ranks and professions should be considered and solved as quickly as possible. Some ideas are suggested to start the exchange of opinions concerning the reasons and consequences observed in the field of practical teaching and teaching materials, Student's Course Book and Maritime –related word stock, in particular. We hope this discussion will make possible to analyse all aspects of Maritime English in use and to draw conclusions about the neccessity of taking them into account when developing the basics of Maritime English linguistic and methodological concepts.

2 LINGUISTIC ASPECT OF MARITIME ENGLISH TRAINING

Researchers emphasyse on the global nature of English (Global English, International English), call English *lingua franca* for the people who work in multinational surrounding. Maritime English is considered to be an operational and working language, the language with some restrictions if the functional characteristics are concerned in the specific area of merchant marine transportations (Ziarati & Ziarati & Calbas & Moussley 2008). The linguistic analysis indicates the availability of considerable lexical "burden" of special terms, quite a short list of grammar structures, strikingly serious set of phonetic peculiarities in Maritime English use. Specific features of ME cause certain difficulties in mastering the system of maritime terms in which a term is not only a language unit; but represents a notion belonging to the special sphere of knowledge. In case all these linguistic factors are taken into account in university curricula, it is possible to foresee that they have a chance of being successfully used for effective training of would-be deck and engineering crew members. Still, as many professionals think, Maritime English is not the whole English language which is required for communication in different spheres of life. The idea of co-relation of General English and Maritime English comes around when developing various teaching/learning materials (study books, tests for self-assessment) where special terms are also considered as assets both of Maritime English and specialism.

3 WORD-STOCK TEACHING/LEARNING MATERIALS AS A PART OF PRACTICAL LEXICOGRAPHY AND TERMINOGRAPHY

Being different from lexicography which is mostly descriptive, terminography is primarily prescriptive; terminography deals with concepts and terms and not with linguistic signs. Terminography is qualified as a synchronic research, uses only experts as informants and is entirely based on systematic classification, which is not the case with lexicography. When discussing the essence of terminology and terminography, it's advisable to take into account that they mainly imply the three basic ideas:

- the set of practises and methods used for the collection, description and presentation of terms;
- a theory for explaining the relationships between concepts and terms;
- a vocabulary of a special subject field (Bergenholtz & Kaufmann 1997).

The present study is based on the teaching materials collected in Introductory Maritime English Course (Student's Book 1 and Student's Book 2) designed in Kyiv State Maritime Academy, Ukraine for the beginners of the Faculty of Navigation. Maritime-Related Vocabulary (English-Russian and English Ukrainian versions) on the following topics:

1 Introducing Oneself. Filling up personal documents; types of documents; interviews.
2 Letters, numbers, colours. Maritime code words. Times at sea and at shore. Languages, nationalities, flags.
3 Maritime jobs and professions. Functions and duties.
4 Places and locations. Countries, water bodies. Other geographical names. Maps and charts. Longitude, latitude.
5 A ship: dimensions, particulars, parts, structure, functional zones.
6 Types of vessels.
7 Motion and directions: navigation, propulsion, engines.
8 Engineering: types of a vessel's equipment.
9 Running the vessel. The bridge. The engine room.
10 Watches and watchkeeping.
11 SMCP: on-board, external. Orders and commands. VHF radio.
12 Daily routines of the crew members.
13 Weather and climate, weather forecast, disasters.
14 Emergency situations.
15 Safety equipment and its location.
16 Steering, mooring, anchoring. Piloting.
17 Ports and port infrastructure. Administration, customs, sanitary inspection.
18 Navigational aids: buoys and lighthouses.
19 Cargoes: types; loading/discharging operations.
20 Shipping documents (basics).
21 Checking supplies.
22 Incidents and accidents at sea. Injuries. First aid.

"The Introductory Maritime English Course" is intended for the first and second year non-native English-speaking learners who are about to commence their Maritime academic career through a Bachelor Degree in Navigation or Marine Engineering. Three influences behind the development of the study book and as such its contents and the form are taken into consideration. These are the

1 lack (or absence) of professional Maritime experience of the students,
2 lack (or absence) of Maritime English language proficiency,
3 lack of General English language competency.

The course book is supposed to meet the interests and requirements of the future seafarers in a new sphere of knowledge whereby the coordinated work of English language teachers and specialists' teachers is required. The Course fills in the current gaps and adds new necessary requirements by combining English language and Maritime specialist skills with the existing General English language foundations. So, both the Maritime-Related vocabulary and the system of exercises on terminology suggested in the Introductory Maritime English Course have the pragmatic purpose of developing skills of speaking fluently in the professional environment of the potential seafarers.

4 MARITIME-RELATED ENGLISH VOCABULARY: ITS STRUCTURE AND TEACHING GOALS

The vocabulary attached to the Introductory Maritime English Course is a multilingual version (English – Ukrainian – Russian) compiled for two main purposes: a) reference for the meaning and spelling, b) reference for word combinations surrounding the term. It consists of over 1500 maritime-related terms, which are evaluated as the most frequent special words and collocations used in the Introductory Maritime English Course.

Example 1 (Translation) **maritime** — ***укр.*** морський; такий, що має відношення до моря чи океану; ***рос.*** морской; имеющий отношение к морю или океану

Example 2 (Combinability)

maritime

-**maritime** *authorities*
-**maritime** *education*
-**maritime** *education and training*
-**maritime** *institutions*
-**maritime** *students*
-**maritime** *academy*
-**maritime** *university*

Example 3 (Core terms and collocations)
sea, a seaman, a seafarer, a seagoing vessel
to sail, a sail, a sailor
to navigate, navigation, a navigator
nautical (captain, officer, terms, tools, mile)
marine, mariner, submarine
maritime, Maritime English, International Maritime Organisation
ship, ship's Master, shipbuilding (industry)
shipping, shipping company, industry
navy , Merchant Navy, naval, naval fleet
crew, crewing company, crew members
cargo, dry, liquid cargo; cargo vessel
vessel, merchant vessel

Example 4 (Word structure, derivation)

-man	**-er, -or**
radioman	to wipe – a wiper
motorman	to dive – a diver
pumpman	to navigate – a navigator
watchman	to oil – an oiler

Example 5 (Key words for a topic production)
Duties and responsibilities
a rank, a title, a license, power – powers, responsibility, to be in charge of something, to be responsible for somebody or something, to ensure something, to take care of something

Example 6 (Functional word group)
SMCP message markers: Instruction
Advice
Warning
Information
Question
Answer
Request
Intention

Example 7 (Conversational basics)
VHF radio communication
to say, to speak, to communicate, to respond, external, on-board, ship-to-ship, ship-to shore, affirmative, negative, instruction, information, advice, warning, question, answer, request, intention, VHF. How do you read me? Over.

Example 8 (Indications for a picture)

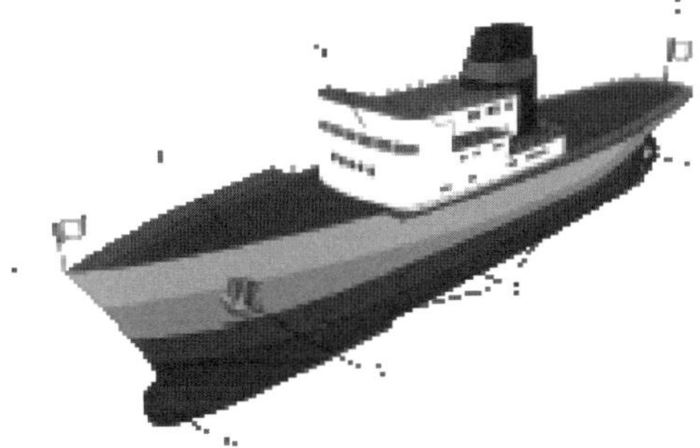

Figure 1. Main parts of a ship

1: Smokestack or Funnel;
2: Stern;
3: Propeller and Rudder;
4: Portside (the right side is known as starboard);
5: Anchor; **6**: Bulbous bow;
7: Bow; **8**: Deck; **9**: Superstructure

Example 9 (Verb and noun groups on a topic)
Mooring. Anchoring.
to fasten, to attach, to reach, to accomplish, to connect, to tie, to pull, to haul, to install, to compare, to resist, to be available;
a rope, mooring lines, a hawser, a hawse pipe, a wire (rope), deck fittings, a bollard, rings, cleats, a cable, a chain, a screw, a shaft, a winch, a capstan, a bitt, snapback, an injury.

Example 10 (Consolidation)
Measurements
to measure-measurements, a dimension, a value, a magnitude, a unit, system of units, to convert-conversion, a circle, a line, a sphere, exact-exactly, approximate-approximately-approximation, a knot, a mile, a nautical mile, Imperial mile, Admiralty mile, a geographical mile, a historical nautical mile, an arc minute, a metre, a kilometer, a degree, a foot-feet, a fathom, a cable, a yard, metrics

Example 11 (Glossary)
a) **The Global Positioning System (GPS) -** a satellite-based navigation system made up of a network of 24 satellites placed into orbit
b) ***Port area*** - a complex of berths, docks and the land where ships and cargoes are served

The above given examples illustrate the opportunities for a language teacher using the maritime-related vocabulary with different insertions including pictures, sets of synonyms, antonyms, collocation, derivation, definition, semantic groups, subject groups, i.e. all possible linguistic tools available.

5 CONCLUSIONS: INTERDISCIPLINARY STRATEGY IN TERMINOLOGY TEACHING

As far as the subject teachers are the source of scientific information about the concept of a nautical term, it should be taken for granted to coordinate the work of language teachers with them. Different levels of conceptual meaning or various definitions of one and the same term may lead to misunderstanding. Compare, for example: *A term* is a word or phrase used as the name of something, especially one connected with a particular type of language.

A term is a word or expression that has a precise meaning in some uses or is peculiar to a science, art, profession, or subject. The role of a language teacher, therefore, is mainly to explain linguistic features and current use of a certain term. "Twinning" as one of the methods to get together in teaching specialist vocabulary and technical terminology may be successful under several conditions: a) high qualification of the two partners trying to achieve one goal for the use of their trainees, b) development of additional teaching materials, for example, "Introduction to the subject" which itself is a time consuming activity, c) design of a syllabus for the classroom activities involving all possible audio and visual aids, d) scrupulous attention to the language (definitions, explanations, discussion, etc. on the particular topic or theme), e) minimization of direct translation as one of the most commonly used methods. Hence, a terminographic research carried out in a certain professional field may become one of the efficient ways of students' language communicative skills stimulation which greatly benefits both parties of the interdisciplinary process.

REFERENCES

Bergenholtz H.& Kaufmann U. 1997. Terminology and Lexicography. A critical survey of dictionaries from a single specialised field. Hermes, *Journal of Linguistics* No.18: 92.

Merriam Webster's Dictionary 2008. Encyclopaedia Britannica

Oxford Advanced Learners Dictionary, 7th Edition. Oxford University Press

Ziarati M.& Ziarati R.&Calbas B.&Moussley L. 2008.

Improving safety at sea by developing standards for Maritime English, *IMLA 16 Proceedings*, 175-181

Piracy Problem

22. Somali Piracy: Relation Between Crew Nationality and a Vessel's Vulnerability to Seajacking

A. Coutroubis & G. Kiourktsoglou
University of Greenwich, London, United Kingdom

ABSTRACT: This paper constitutes an effort to substantiate whether there are certain nationalities of crews which are for ethnic and / or cultural reasons more (or less) vulnerable to fall victims of Pirates off Somalia. Such groups (if there are any) in effect indirectly 'support' Somali piracy and for this reason they are being re-ferred to throughout the paper as "Passively Supportive Crews".
The method (and the rational) in use within this paper is straightforward. Over a three and a half year period (2007 – June/2010) an analysis is being conducted of all the reported (to the I.M.O. and I.M.B.) attacks in the region off Somalia.
The analysis focuses on the crew composition of the attacked vessels with special interest cast upon those Ships (meaning the crews) which eventually succumbed to the pirates and were in the end seajacked.

1 INTRODUCTION (ASSOCIATED PRESS, MARCH 2010)

On Thursday the 4th of March 2010, Somali pirates hit a Spanish fishing boat off the coast of Kenya with a rocket-propelled grenade as private security on board returned fire at the would-be seajackers. The successful defense of the fishing vessel Albacan illustrated two trends driving up the stakes for sailors and pirates off the Horn of Africa:

Better trained and protected crews are increasingly able to repel attacks, but Pirates eager for multi-million-dollar ransoms are now resorting to violence much more often to capture ships.

Two-thirds of attacks by Somali Pirates are being repelled by crews alone, without the aid of the coalition warships that patrol the Gulf of Aden, according to an analysis by the London-based International Maritime Bureau. Most did so without the use of armed guards, although in 2009 private security contractors helped repel pirates in at least five incidents off the Somali coast.

As it gets harder for pirates to capture ships, the Somali gangs are more likely to fire at sailors with automatic weapons in order to force vessels to stop. The IMB states that only seven ships were fired upon worldwide in 2004 but that 114 ships were fired upon in 2009 off the Somali coast alone. That is up from thirty-nine incidents off Somalia and in the Gulf of Aden in 2008.

Most crews now post extra lookouts, register with maritime authorities and practice anti-piracy drills. Increasing speed and maneuvering, so that a ship produces more wake or heads into rough waves, can also make it more difficult for pirates.

The International Maritime Bureau does not recommend using armed guards due to potential legal issues and fears of starting an arms race with the pirates or increasing the danger to crews. Armed guards on ships may encourage pirates to use their weapons more — a prediction that appears to have become reality.

Some ships have been forced to rely on sailors' ingenuity. Crews have thrown everything from oil drums to wooden planks at would-be seajackers clambering up ladders. In 2009, a crew played the sound of dogs barking over an amplifier to frighten off attackers.

Better training and preparation means that although 2009 saw 217 Somali pirate attacks — the highest number on record — most were unsuccessful. Forty-seven ships were taken, about the same as in 2008, which saw 111 attacks, according to the International Maritime Bureau.

The attacks are becoming more dangerous for crew members though. In 2009, more than twenty ships were fired upon with rocket-propelled grenades, including tankers and chemical tankers. In one incident, two grenades lodged in the door of a ship's bridge — the area where the captain steers from. Many other ships were damaged by small-arms fire, according to reports from IMB.

Four sailors died and ten were injured off Somalia in 2009. Two were killed during rescue attempts

— one by Yemeni forces and one by the French — and another died in captivity. The fourth was killed by a bullet during the attack.

In 2009, the average ransom was around $2 million, giving the pirates a total haul of around $100 million during that year. According to industry officials just up to April 2010, two ransoms paid were around $3 million and $7 million.

As an industry analyst wryly puts it: "There's a commercial calculation as well as a humanitarian one..... It's cheaper to pay a bit more a bit more quickly than a bit less over a longer period of time, because of associated costs like compensation to the sailors, lost work time, and possibly a loss in the value of the cargo."

As we are still tackling piracy in accordance with the 'International Law in Time of Peace', it is a matter of cooperation between the various stakeholders. It is in this goal that Private (Vessel Owners, Ship Management Companies), National (Flag States, Port States) and Supranational (UN, IMO, EU etc) interests and objectives should converge.

Unfortunately many believe that 'off-the-shelf' solutions like barbed wire, high pressure water hoses or even armed guards on board vessels can on their own effectively counter the piracy scourge. This is a fallacy and a very costly-one if not fatal. Only cooperation among all kinds of relevant authorities / market players can create the right environment for Maritime Security to come to fruition.

All in all, as the Athenian philosopher Socrates put it squarely right some 2,500 years ago: "The Man is the Ultimate measure of Everything....."

1.1 *Literature review (Gekara 2008)*

Although the forces of economic globalization have greatly diminished national economic barriers in the past four decades, labour is yet to enjoy the same global mobility that capital and finance enjoy. In the main, labour continues to be locally and nationally organized and the state still wields immense regulatory control through immigration restrictions across borders (Holton, 1998). Other obstacles like cultural, and language barriers, and variations in the education, training and qualification systems of different countries also restrict the international movement of labour (Lauder and Brown, 2006).

However, in shipping, the growth of the Global Labour Market for seafarers has significantly increased the mobility of seafarers in the past few years (Wu, 2004). Furthermore, the mobile nature of seafaring employment, combined with the international harmonization of training and certification in the profession and the use of English as the accepted international language of seafaring, defines seafaring in distinctive ways.

Ship-owners have, over the years, designed crewing policies which enable them to increase their competitive advantage in terms of cost effectiveness. These policies direct their recruitment strategies and have, over the years, resulted in increasing the prevalence of seafarers from low-wage developing countries.

The worldwide supply of seafarers in 2005 was estimated to be 466,000 officers and 721,000 ratings (BIMCO / ISF, 2005). The OECD countries (North America, Western Europe, Japan etc.) remained an important source of officers, although Eastern Europe has become increasingly significant with a large increase in officer numbers. The Far East and South East Asia (the "Far East"), and the Indian subcontinent remain the largest sources of supply of ratings and are rapidly becoming a key source of officers.

On the other hand, the 2005 estimate of worldwide demand for seafarers was 476,000 officers and 586,000 ratings.

2 SOMALI PIRACY (INTERNATIONAL MARITIME BUREAU, 2007-2010 Q2 REPORTS)

2.1 *Review of the recent past (January 2010 – June 2010)*

Somali pirates attack vessels in and around the following areas:

Coasts along the northern, eastern and southern Somalia;

- Red and Arabian Seas;
- Western Indian Ocean (more than 1,000 nm away from the eastern Africa basin);
- Gulf of Aden;
- Seas off the coasts of Kenya, Tanzania. Seychelles, Madagascar and Oman;
- Straits of Bab el Mandeb.

From January to June 2010, there have been reports of 100 incidents carried out by suspected Somali pirates. The incidents varied in geographical location encompassing the waters already mentioned above. A total of 544 crew members have been taken hostage and a further 10 have been injured. There have been 51 attacks off the East and South coasts of Somalia, another 33 attacks in the Gulf of Aden, 14 attacks in the Southern Red Sea, 2 reported in the Arabian Sea. 27 vessels have been reported seajacked in this period.

As of the 30th of June 2010, suspected Somali pirates held 18 vessels for ransom with 360 crew members of various nationalities as hostages.

Somali pirates attack all kinds of vessels: General Cargo, Bulk Carriers, Tankers, Ro-Ro, Liners, Fishing vessels, Sailing Yachts and Tugboats.

The piratical activities peak each year from September until April and then their numbers start to drop due to the monsoons that prevail in the area. On a 24 hr per day analysis basis, the most dangerous periods for piratical attacks are the dusk and the daybreak.

Table 1, Vessel Seajacks off Somalia from January 2007 till June 2010

INDE	SHIP NAME	VESSEL TYPE	MONTI	YEA
1	ROZEN	GENERAL CARGO	February	2007
2	DANICA WHITE	GENERAL CARGO	June	2007
3	GOLDEN NORI	CHEMICAL TANKER	October	2007
4	SVITZER KORSAKOV	TUG	February	2008
5	LE PONANT	PASSENGER VESSEL	April	2008
6	AMIYA SCAN	GENERAL CARGO	May	2008
7	LEHMAN TIMBER	GENERAL CARGO	May	2008
8	YENEGOA OCEAN	TUG	August	2008
9	IRENE	CHEMICAL TANKER	August	2008
10	BBC TRINIDAD	GENERAL CARGO	August	2008
11	AL MANSURAH	GENERAL CARGO	September	2008
12	GENIUS	TANKER	September	2008
13	CENTAURI	BULK CARRIER	September	2008
14	CAPTAIN STEFANOS	BULK CARRIER	September	2008
15	CARRE D' AS IV	YACHT	September	2008
16	STOLT VALOR	TANKER	September	2008
17	GREAT CREATION	BULK CARRIER	September	2008
18	WAEL H.	GENERAL CARGO	October	2008
19	ACTION	CHEMICAL TANKER	October	2008
20	AFRICA SANDERLING	BULK CARRIER	October	2008
21	YASA NESHLIHAN	BULK CARRIER	October	2008
22	CHEMSTAR VENUS	TANKER	November	2008
23	SIRIUS STAR	TANKER	November	2008
24	BISCAGLIA	CHEMICAL TANKER	November	2008
25	CEC FUTURE	GENERAL CARGO	November	2008
26	DELIGHT	BULK CARRIER	November	2008
27	TIANYU N. 8	FISHING TRAWLER	November	2008
28	KARAGOL	CHEMICAL TANKER	November	2008
29	BOSPHORUS PRODIGY	GENERAL CARGO	December	2008
30	LONGCHAMP	TANKER	Janurary	2009
31	SALDANHA	BULK CARRIER	February	2009
32	NIPAYIA	CHEMICAL TANKER	March	2009
33	TITAN	BULK CARRIER	March	2009
34	BOW ASIR	CHEMICAL TANKER	March	2009
35	MALASPINA CASTLE	GENERAL CARGO	April	2009
36	PATRIOT	BULK CARRIER	April	2009
37	BUCCANEER	TUG	April	2009
38	MAERSK ALABAMA	CONTAINER VESSEL	April	2009
39	TANIT	YACHT	April	2009
40	IRENE E.M.	BULK CARRIER	April	2009
41	HANSA STAVANGER	CONTAINER VESSEL	April	2009
42	ALMEZAAN	GENERAL CARGO	May	2009
43	ARIANA	BULK CARRIER	May	2009
44	VICTORIA	GENERAL CARGO	May	2009
45	MARATHON	GENERAL CARGO	May	2009
46	CHARELLE	GENERAL CARGO	June	2009
47	HORIZON 1	BULK CARRIER	July	2009
48	AL KHALIQ	BULK CARRIER	October	2009
49	LYNN RIVAL	YACHT	October	2009
50	KOTA WAJAR	CONTAINER VESSEL	October	2009
51	DELVINA	BULK CARRIER	November	2009
52	FILITSA	BULK CARRIER	November	2009
53	MAERSK ALABAMA	CONTAINER VESSEL	November	2009
54	MARAN CENTAURUS	TANKER	November	2009
55	NAVIOS APOLLON	BULK CARRIER	December	2009
56	ST. JAMES PARK	CHEMICAL TANKER	December	2009
57	PRAMONI	CHEMICAL TANKER	January	2010
58	ASIAN GLORY	VEHICLE CARRIER	January	2010
59	RIM	GENERAL CARGO	February	2010
60	AL NISHR AL SAUDI	TANKER	March	2010
61	UBT OCEAN	CHEMICAL TANKER	March	2010
62	FRIGIA	BULK CARRIER	March	2010
63	TALCA	REEFER	March	2010
64	ICEBERG I	RO-RO	March	2010
65	VISHVA KALYAN VRL	DHOW	April	2010
66	JIH CHUN TSAI	FISHING TRAWLER	April	2010
67	SAMHO DREAM	VEHICLE CARRIER	April	2010
68	RAK AFRIKANA	GENERAL CARGO	April	2010
69	PRANTALAY 11	FISHING TRAWLER	April	2010
70	PRANTALAY 12	FISHING TRAWLER	April	2010
71	PRANTALAY 14	FISHING TRAWLER	April	2010
72	VOC DAISY	BULK CARRIER	April	2010
73	AL-ASA'A	DHOW	May	2010
74	MOSCOW UNIVERSITY	TANKER	May	2010
75	TAI YUAN No 227	FISHING TRAWLER	May	2010
76	MARIDA MARGUERITE	CHEMICAL TANKER	May	2010
77	PANEGA	CHEMICAL TANKER	May	2010
78	ELENI P	BULK CARRIER	May	2010
79	AL JAWAT	DHOW	June	2010
80	QSM DUBAI	GENERAL CARGO	June	2010
81	GOLDEN BLESSING	CHEMICAL TANKER	June	2010

Source: International Maritime Organisation, I.M.O., Monthly Reports on acts of piracy and armed robbery against ships from January 2007 till June 2010

Over the years the Somali pirates have evolved in their use of weapons and tactics. Currently they are using automatic rifles and rocket propelled grenades (RPGs). They have also advanced from using dilapidated fishing boats to launch their attacks into using large pirated trawlers as mother-ships to support smaller attack units.

2.2 *Cumulative picture (January 2007 – June 2010*

For the purpose of the present analysis a compilation has been created of all the successful vessel seajacks off Somalia (Table 1).

The compilation includes the vessel's name, her type, flag, gross tonnage, the date of the seajack, but above all the break-down of her crew in terms of nationalities. In total 81 Seajackings have been recorded from January 2007 until June 2010 and they feature a great variety in terms of vessel types, registries, gross tonnage etc.

Based on the compilation a matrix was produced on the crew nationalities of the vessels which eventually succumbed to the Somali Pirates and they were taken to captivity (Table 2).

Table 2, Break Down of Seajacked Vessels Crew Nationalities (Vessel Seajacks off Somalia from 2007 till June 2010)

INDEX	COUNTRY	INCIDENT LEVELS	2003 INTERN. LEVELS	Number of Seamen
1	Philippines	26,58%	27,80%	392
2	India	9,02%	6,60%	133
3	China	6,58%	6,10%	97
4	Turkey	5,56%	4,00%	82
5	Russia	5,49%	7,00%	81
6	Ukraine	5,29%	6,40%	78
7	Thailand	5,22%		77
8	Shri Lanka	3,73%		55
9	Romania	3,05%		45
10	Bulgaria	2,98%		44
11	U.S.A.	2,85%		42
12	Burma	2,78%	2,20%	41
13	Indonesia	2,51%	3,50%	37
14	France	1,97%		29
15	Syria	1,83%		27
16	Egypt	1,69%		25
17	Poland	1,42%	3,00%	21
18	Georgia	1,29%		19
19	Greece	1,15%	2,80%	17
20	Kenya	0,88%		13
21	S. Korea	0,81%		12
22	Nigeria	0,75%		11
23	Tuvalu	0,75%		11
24	Yemen	0,75%		11
25	Italia	0,68%		10
26	Pakistan	0,54%		8
27	Vietnam	0,54%		8
28	Iran	0,47%		7
29	Bangladesh	0,34%		5
30	Danemark	0,34%		5
31	Germany	0,34%		5
32	U.K.	0,34%		5
33	Croatia	0,14%		2
34	Ghana	0,14%		2
35	Mozambique	0,14%		2
36	Singapore	0,14%		2
37	Somalia	0,14%		2
38	Taiwan	0,14%		2
39	Cameroon	0,07%		1
40	Esthonia	0,07%		1
41	Fiji	0,07%		1
42	Hong Kong	0,07%		1
43	Ireland	0,07%		1
44	Japan	0,07%		1
45	Lithuania	0,07%		1
46	S. Arabia	0,07%		1
47	Serbia	0,07%		1
48	Slovakia	0,07%		1
	Total	100%		1.475

Source: Various Internet based Press Reports

It seems that mainly the citizens of the Philippines (26.58%), India (9.02%), China (6.58%), Turkey (5.56%), Russia (5.49%), Ukraine (5.29%) and Thailand (5.22%) bore the brunt of Somali Piracy.

The incident compilation also enabled the production of the phenomenon's statistical profile in terms both of the vessel type (Table 3) and registry (Table 4).

Table 3, Types of Vessels successfully Seajacked off Somalia from 2007 till June 2010

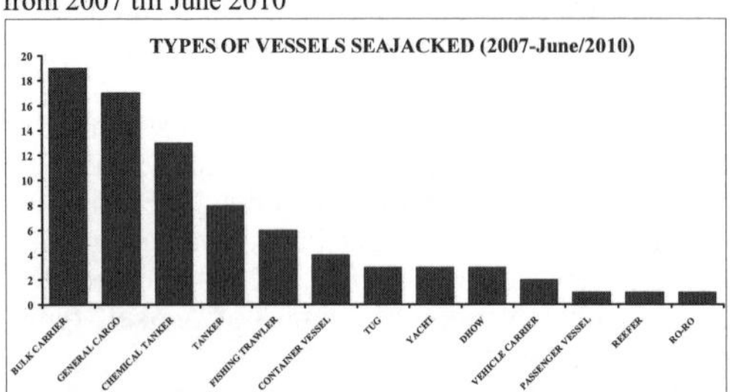

Source: International Maritime Organisation, I.M.O., Monthly Reports on acts of piracy and armed robbery against ships from January 2007 till June 2010

Table 4, Flags of Vessels successfully Seajacked off Somalia from 2007 till June 2010

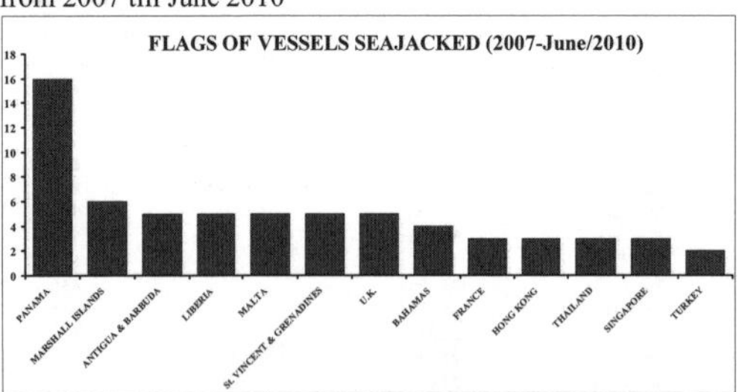

Source: International Maritime Organisation, I.M.O., Monthly Reports on acts of piracy and armed robbery against ships from January 2007 till June 2010

In this case, it seems that Somali Pirates have shown preference both for bulk carriers (23.4% of all seajacks within the study's timeframe) and the registry of Panama (20.2%). Both findings are statistically in line with the industry's "ground realities" since bulk carriers represent 35% of the international high-seas commercial fleet and the Registry of Panama is by far the largest worldwide, with 14% of the International Fleet under its flag.

3 COMPARATIVE STATISTICS ON CREWS OF SEAJACKED VESSELS

In 2003 the Seafarers International Research Centre (S.I.R.C.) of Cardiff University published its most recent report / survey on "*The Global Labour Market for Seafarers Working aboard Merchant Cargo Ships*" (The Global Labour Market for Seafarers Working aboard Merchant Cargo Ships, 2003).

Philippines were found to dominate the global seafarer labour market with 28% of the sample studied holding Filipino nationality. Russians, Indians, Ukrainians, and Chinese nationals all constituted a similar proportion of the sample (between 6% and 7%) followed by Turkey, Indonesia, Poland, Greece and Myanmar in descending order (Tables 5A & 5B).

Table 5A, % of Nationals in Crews of Seajacked Vessels

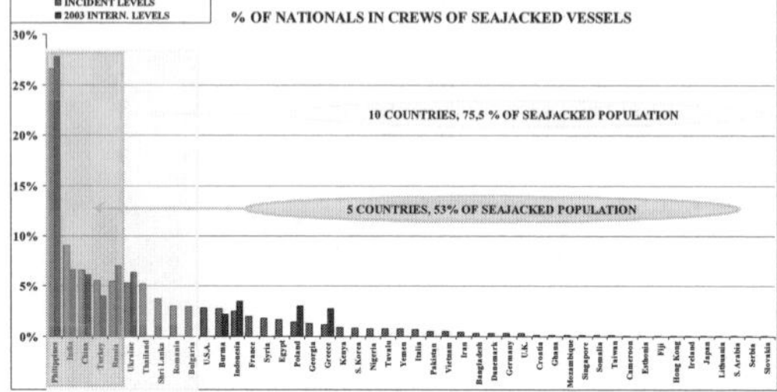

Source: The Global Labour Market for Seafarers Working Aboard Merchant Cargo Ships (2003) and Various Internet based Press Reports

Table 5B, % of Nationals in Crews of Seajacked Vessels

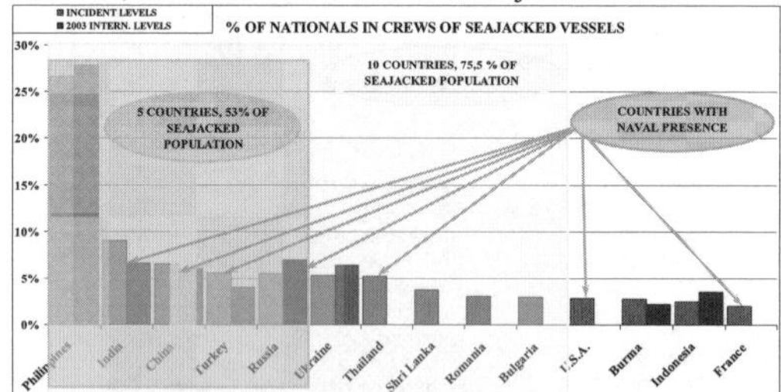

Source: The Global Labour Market for Seafarers Working Aboard Merchant Cargo Ships (2003) and Various Internet based Press Reports

These ten nationalities constitute 70% of the total sample.

By far the largest group of ratings by nationality is Filipino. Filipino seafarers constitute more than a third of all ratings. Their domination of the ratings labour market is significant and all of the other nationalities, even in the top ten represented amongst ratings, can be considered to represent minor groupings by comparison.

Whilst seafarers from the Philippines dominate the labour market overall, their domination (compared with other nationalities) is less marked with regard to senior officer positions. They remain the largest nationality group (both in absolute and relative terms) amongst senior officers; however nationalities are much more evenly distributed in the senior officer category than they are in general. Filipinos constituting roughly 11% of senior officers are closely followed by Russians who account for almost 10% of senior officers. Ukrainians, Greeks, and Indians account for approximately 6-7% of senior officers each, and Chinese, Polish, South Korean, German and Turkish officers are all represented at the level of around 4% (each). There is a greater variety of nationalities represented at senior officer level than there is across the board.

Amongst junior officers the domination of the labour market by Filipinos appears as a marked feature. 24% of junior officers were found to be of Filipino nationality and this proportion is considerably larger than the one featured by the second largest national group, Russians, who made up approximately 9% of the sample. Indian, Ukrainian, and Chinese nationals constitute between around 7% and 8% of the sample (each), with Polish, South Korean, Indonesian, and Romanian seafarers constituting smaller groups amongst the top ten nationalities of junior officer. Ceteris paribus, this distribution of junior officers suggests that in the future Filipinos will constitute a much larger proportion of senior officers across the global fleet. However, should there be any barriers to the transition of Filipino seafarers from junior to senior officer status; these figures could suggest that there may be problems in later years for companies wishing to recruit senior officers.

4 CONCLUSIONS

Analysts around the world have focused on the proximate cause(s) of Somali Piracy and many have come to silently believe (if not publicly suggest) that at least in some cases the Somali Pirates enjoy help from the inside. As mentioned in the introductory paragraph, the purpose of this paper is to investigate potential links between the crew nationalities of seajacked vessels and the occurrence of the seajacks themselves.

Within this study, a comparison was undertaken between the crew nationalities of seajacked vessels against the overall composition of crew nationalities of the global mariner population. Although the latter profile is fairly outdated (most recently produced in 2003) it can still provide a good insight into the status quo in terms of the nationalities of the international shipping crews.

The analysis performed bore no proof whatsoever that there are 'passively supportive crews' of Somali Piracy. The breakdown by nationality of the crews falling victim of piracy is broadly (and within the statistical error-area of ±3%) in line with the overall participation of crew nationalities in international shipping.

In more detail (Tables 5A & 5B):

- Filipinos represent 27.8% of the international seafarer population and 26.6% of the seajacked crews.
- Correspondingly, Indians represent 6.6% of the seafarer population and 9% of seajacked crews.
- Last but not least, the Chinese nationals feature almost an "utter balance" within the two groups with 6.1% and 6.6% respectively.

This conclusion does not imply that the crew composition and the training are not factors of value to be considered when combating piracy. It simply suggests that the crew nationality does not appear as an "operational driver" in the case of successful seajacks.

Amongst secondary observations the following ones stand out conspicuously:

1. The five nations (Philippines, India, China, Turkey and Russia) that provide international shipping with more than half of its seafarers (51.5%) bear (through their nationals - seafarers) the main brunt (53.22%) of seajacks off the coast of Somalia.
2. Among 48 countries in the "seajacked" crew population from 2007 until June 2010, 3 out of 4 seafarers are nationals of 10 countries (Philippines, India, China, Turkey, Russia, Ukraine, Thailand, Sri-Lanka, Romania and Bulgaria).
3. It seems that the presence of a country's Navy (India, China, Turkey and Russia) off Eastern Africa has no impact whatsoever on the number of its nationals that fall victims of Somali seajacks.
4. A remarkable observation though demands some extra attention:
5. Although more than one out of four seafarers employed onboard seajacked vessels is a Filipino, this island country and indeed maritime nation has no naval presence off Somalia.

The commercial impact of piracy on the shipping industry has been massive with more than $100 million paid as ransom worldwide in 2009 alone. The insurance premium for passage through the Gulf of Aden has increased 10-fold and continues to increase further fuelling the 'kidnap-for-ransom' marine insurance industry. Every ship operating firm can always re-route its voyages via the Cape of Good Hope, a decision though that eventually entails higher cost(s) for the cargo movement. This diversion also affects the delivery - times of commodities worldwide.

The shipping industry (and global trade on the whole) has been badly affected and international authorities and governments should take a different and more aggressive approach to end the pirate menace off Somalia. Given the recent post 9/11 example, declaring war against pirates appears the logical step after this long period of international procrastination. It is worth remembering that the Golden Age of Piracy came to an end shortly after the signing of the Treaty of Utrecht in 1713 when, in effect, the European states declared war on piracy and sent their fleets after them. The results were remarkably immediate and the rest became history.

In such a dire situation, it is of no surprise for plot-scenarios to come to the surface and to suggest that there may be some "operational drivers" (nationalities of crews among other things) behind the spike of seajack cases off Somalia.

This paper did not aim to exonerate a certain professional group let alone to extol an existing practice. It just intended to shed a glimmer of light on a scientifically "uncharted territory" and consequently investigate a potential nexus between seajacked-vessel's crew nationalities and the seajack itself.

REFERENCES

Associated Press, Arms race on high seas: Armed pirate attacks soar. [WWW] < URL: http://www.ap.org/> [Accessed 5 March 2010]).

BIMCO/ISF. 2005. Manpower Update: The World-wide Demand for and Supply of

Seafarers. The University of Warwick: Institute for Employment Research

Holton, R. 1998. Globalization and the Nation-State. London: Mcmillan Press Ltd.

International Maritime Bureau (2007 - 2010 (Q2) Reports). Report on Piracy and Armed Robbery against Ships.

International Maritime Organisation, I.M.O.. Monthly Reports on acts of piracy and armed robbery against ships (January 2007 till June 2010)

Lauder, H., Brown, P., Dillabough, J. A. and Halsey, A. S. (eds) 2006. Education, Globalization and Social Change. Oxford: Oxford University Press.

Seafarers International Research Center, Cardiff University, (2003). The Global Labour Market for Seafarers Working Aboard Merchant Cargo Ships. [WWW] < URL: http://www.sirc.cf.ac.uk> [Accessed 30 August 2010]).

United Nations Conference on Trade and Development, U.N.C.T.A.D, (2009). Review of Maritime Transport. www <URL: http://www.unctad.org/en/docs/rmt2009ch2_en.pdf> [Accessed 30 August 2010]).

Victor Oyaro Gekara, Globalization, State Strategies and Shipping Labour Market, (January, 2008). Seafarers International Research Centre (SIRC). [www] < URL: http://www.sirc.cf.ac.uk > [Accessed 7 May 2010])..

Wu, B. 2004. Participation in the Global Labour Market: Experience and Responses of

Chinese Seafarers. Maritime Policy and Management 31(1), pp. 69 – 82.

23. Influence of Pirates' Activities on Maritime Transport in the Gulf of Aden Region

D. Duda & K. Wardin
Polish Naval University, Gdynia, Poland

ABSTRACT: Modern piracy is one of the items appearing on the seas, which has a great impact on maritime transport in many regions of the world. Changes that happened at the end of XX and beginning of XXI century became significant in the renaissance of piracy. The problem is present in many parts of the world but it become a real threat in year 2008 around a small country of Somalia and in the area called the Horn of Africa especially in the region of Gulf of Aden. Because international waters are very important for maritime transport so pirates' attacks have great influence over this transport and on international community.

1 PIRACY – DEFINITION AND MAIN AREAS OF PIRATES' ACTIVITIES

Piracy is an activity known and grown for thousands of years. At present in many parts of the world it is treated as a type of legacy or rather part of tradition and so also gladly continued by the population who is experiencing poverty and hunger. Modern day pirates are particularly active in the regions in the waters of the intensive transport by sea. Piracy for many years was treated as an individual problem in each country the coast which existed, and it was not considered as a serious threat to a maritime transport. Such an approach of communities and international institutions to this issue caused the negation of this problem and treating the difficulty as not the most important one. Looking at the world in terms of maritime transport and its more than 95% of the share in the general transport, and also 80% share in the overall transport of crude oil, petroleum and its derivatives, this issue should be put on the first place[1]. The lack of the activities caused that the problem has not disappeared but it has arisen at the end of XX and the beginning of XXI century and has become an immense difficulty for the maritime transportation in many parts of the world.

While talking about piracy it should be clear what is understood by this phrase. The easiest and most understandable definition of piracy is given by International Maritime Bureau (IMB) and according to IMB piracy is defined as*: an act of boarding or attempting to board any ship with the intent to commit theft or any other crime and with the intent or capability to use force in the furtherance of that act*[2].

As mentioned before, the problem is not equally the same in all places where piracy flourishes in the XXI century. Generally speaking we can distinguish five most dangerous regions in the world, as figure 1 shows below, which are really infected with pirates' activities and it influences
maritime transportation in a great matter. These are the following:

- Western and eastern coasts of Africa and the Red Sea;
- The Horn of Arica and the Gulf of Aden;
- The coast of south-east Asia and northern coasts of the Indian Ocean;
- The coast of south America;
- The coast of the Gulf of Mexico.

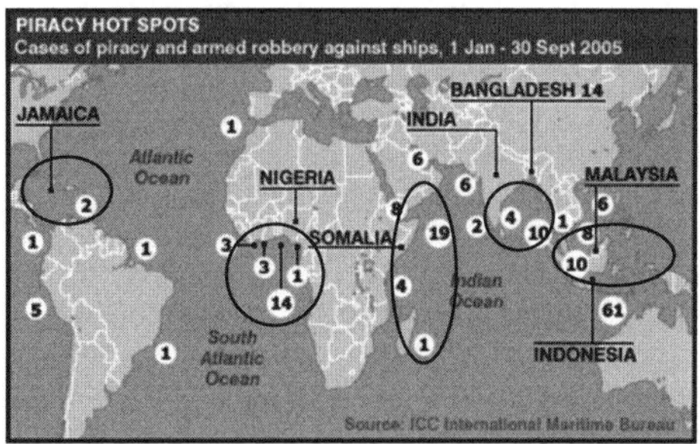

Figure 1, Piracy hot spots, http://www.southchinasea.org/docs/Threats.pdf, 28.12.2008.

[1] Wardin K., „Ocena zagrożeń bałtyckich strumieni transportowych działaniami terrorystycznymi", 25, Belstudio Warszawa 2007.

[2] Hansen S. J., *Piracy in the greater Gulf of Aden,* Norwegian Institute for Urban and Regional Research, 3, London 2009.

The article focuses only on one but very significant region, the Gulf of Aden and the Somali Basin, although the problem is very extensive and present in other regions as well. The Gulf of Aden and the Somali Basin is strictly connected with Somalia, a country situated in the Horn of Africa, and problems related to this country. To understand all aspects of piracy there it is necessary to learn briefly about Somalia as an African country.

2 SOMALIA AND ITS WATERS – A PIRATE-INFESTED COUNTRY

Somalia has made international headlines for almost two decades, first as a place of civil war characterized by clan warfare and humanitarian catastrophe, then as a failed state, and finally as source of modern piracy. Somalia has been without an effective central government since 1991. In that year President Barre was overthrown by opposing clans. But they failed to agree on a replacement and plunged the country into lawlessness and clan warfare. Years of fighting between rival warlords and an inability to deal with famine and disease have led to the deaths of up to one million people. After the collapse of the Siad Barre regime in 1991, the north-west part of Somalia unilaterally declared itself the independent Republic of Somaliland. The territory, whose independence is not recognised by international bodies, has enjoyed relative stability[3]. A two-year peace process, led by the Government of Kenya under the auspices of the Intergovernmental Authority on Development (IGAD), concluded in October 2004 with the election of Abdullahi YUSUF Ahmed as President of the Transitional Federal Government (TFG) of Somalia and the formation of an interim government, known as the Somalia Transitional Federal Institutions (TFIs). President YUSUF resigned late in 2008 while United Nations-sponsored talks between the TFG and the opposition Alliance for the Re-Liberation of Somalia (ARS) were underway in Djibouti. In January 2009, following the creation of a TFG-ARS unity government, Ethiopian military forces, which had entered Somalia in December 2006 to support the TFG in the face of advances by the opposition Islamic Courts Union (ICU), withdrew from the country. The TFIs are based on the Transitional Federal Charter (TFC), which outlines a five-year mandate leading to the establishment of a new Somali constitution and a transition to a representative government following national elections. However, in January 2009 the TFA amended the TFC to extend TFG's mandate until 2011. While its institutions remain weak, the TFG continues to reach out to Somali stakeholders and to work with international donors to help build the governance capacity of the TFIs and to work toward national elections in 2011[4].

Somaliland is not the only part of the country which declared independence and does not want to be ruled by federal government[5]. For the situation has not changed in the country and the people were starving to death, they turned into piracy considering it as a 'modern way of living' in such difficult times.

During August 2008, the frequency of Somali piracy exploded and the drastic increase in occurrence meant that waters adjacent to Somalia became the most pirate-infested waters in the world[6]. However, Somalia as a country is not pirate infested, the pirates usually operate out of only several regions, using only certain ports to anchor their hijacked ships. In order for piracy to occur there must be available targets - sea traffic in the area where potential pirates might operate. This is probably the most obvious reason why the Gulf of Aden and the Somali Basin is a very profitable region to practice piracy.

The Gulf of Aden[7], and waters around Somalia are important areas for navigation. The Gulf of Aden is located between the north coast of Somalia and the Arabic Peninsula and connects the Indian Ocean through the Strait of Bab el-Mandeb with the Red Sea. This is the trail which traverses approximately 21 thousand vessels annually, transporting goods and production of crude oil from the Persian Gulf to Europe and North America. In Arabic, Bab el-Mandeb means "gate of tears", referring to the exceptionally difficult navigation in the Strait. Its length is about 50 km and the width at the narrowest point about 26 km. There is Perim island situated in the middle, which divides it into two parts Bab Iskandar (Strait of Alexander) and Dact al-Majun. The waters on the whole width are territorial waters of the coastal States (Yemen, Djibouti) and the shipping takes place on the basis of the law for the transition of the transit. In 2007 3,3 million barrels of crude oil were transported this way per day out of a world total of about 43 million barrels per day, mainly to Europe, the United States and Asia. The waterway is part of the important Suez Canal shipping route between the Mediterranean Sea and the Arabian Sea in the Indian Ocean. The gulf is known by the nickname 'Pirate Alley' due to the large

---

[3] *Somalia country profile*, http://news.bbc.co.uk/2/hi/africa/country_profiles/1072592.stm, 16.11.2010.

[4] The world factbook: Somalia, https://www.cia.gov/library/publications/the-world-factbook/geos/so.html, 22.11.2010

[5] The other parts: Xizbul Islam, Hrakat al-Shabab Mujahideen, Unaligned or Neutral, other countries. *Somalia*, http://en.wikipedia.org/wiki/Somalia, 16.11.2010.

[6] Reports on Acts of Piracy and Armed Robbery against ships, Annual Report 2008, International Maritime Organization (2009) MSC.4/Circ.115

[7] Its area - 259,000 km², average depth 1359 m, maximum depth 5390 m (Alula Fartak trench).

amount of pirate activity in the area. The Strait of Bab el-Mandeb separates the Gulf of Aden and the Red Sea, and both coasts are occupied by soldiers. In particular, the African coast, which was a witness in the past to the border disputes and resulted in numerous posts to keep the area as safe as possible. Therefore, the Strait itself is not visited by pirates so often.

The figures show definitely that the route is very important for maritime transportation of oil but not only, and so the safety of this region should be the priority for international community. The described route is the shortest sea way to Europe and North America, allowing to save an average of 6,000 nautical miles and a journey around the Cape of Good Hope, which significantly reduces the time of transport and fuel consumption. It should be also added that, due to both the width and depth, which restricts the movement of the units in the Suez Canal, some of the vessels must travel around Africa to get on the Mediterranean and to Americas. It is mainly about super tankers called VLCC (very large crude carriers).

3 POSSIBLE FACTORS RESPONSIBLE FOR PIRACY IN SOMALIA

According to some scientists, observing the piracy in the area, there are several factors to be taken under consideration while analysing the problem: culture, exclusion and relative deprivation, poverty, organizational sponsorship, failure of legal and maritime counter-strategies, and weak/weakening state/institutional structures[8].

These six factors can be collected in two sets of possible reasons for piracy. The first one would be: poverty, organizational sponsorship, failure of counter-strategies and weak/weakening state/institutional structures, which tends to view piracy as a product of rational cost-benefit analyses conducted by the potential pirates. Basically, it is claimed that people engage in piracy because they benefit more from it than from other, alternative activities, either because there are no alternatives (lack of work opportunities), or because the benefits that can be achieved by piracy are really great. Piracy exists there rather as the result of a balance between expected gains from piracy, and expected losses from working as pirates[9].This could be due to several reasons. Punishment for piracy could be weak, because the state or institutions are so weak that piracy cannot be punished, which is true in case of Somalia. The government might not want to fight piracy because of good illegal profits from the trade or because confusion in legal matters acts as a hindrance to punishment.

[8] Hansen S. J., op.cit., 4-5.

[9] Ibidem, 7.

Another two factors: culture, exclusion and relative deprivation focuses on different matters. Culture could lead to some kind of social legitimacy of piracy. In the case of Somalia, the piracy traditions are weak, and thus lack the power to explain the relatively modern phenomena of piracy. The relation between those two ideas, culture and tradition of piracy is rather poor in case of Somalia so this reason cannot be a real explanation in this case. Another reason for piracy in this country, related to exclusion and relative deprivation is so called the ‘Coast Guard’ version claimed by the pirates themselves focuses on piracy as a product of the need to prevent illegal fishing. Pirates are kind of coast guards patrolling and protecting Somali waters from illegal fishing. The next idea is connected with another version of the ‘empty sea’. The argument is linked to the poverty of the country and the cost/benefit balance. It claims that the pirates simply have no alternatives due to overfishing the sea is said to have become empty.

There is also the third version suggesting that piracy started out as a defensive measure taken due to illegal foreign fishing, which over time has turned into professional piracy. Although as given above the reasons can differ there still is one the most important factor in motivating pirates to engage in this activity and it is the profit. This is the motive that appears in almost every interview with an arrested pirate. No matter what is or are real reasons for piracy in this region the fact is that the frequency of attacks has dramatically increased in 2008.

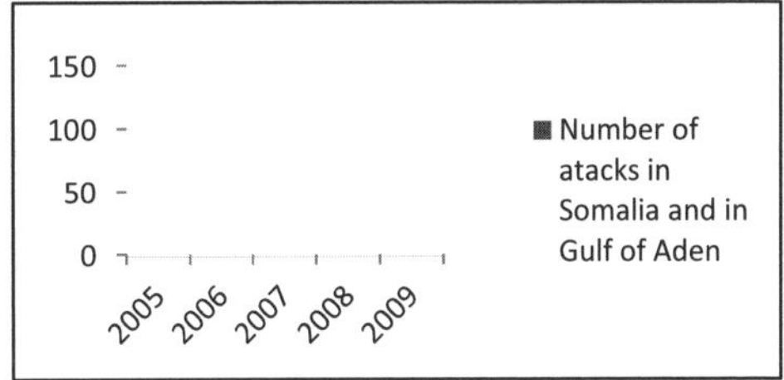

Figure 2, Number of pirate attacks in Somalia and in Gulf of Aden, source: Reports on Acts of Piracy and Armed Robbery against ships, Annual Report 2008, International Maritime Organization (2009) MSC.4/Circ.

The most warring is the fact that in 2010, according to Best Management Practices book 3 (BMP3[10]) the is a significant increase in range of pirates attacks. The high-risk area defined in BMP3 has been expanded beyond the Gulf of Aden to the area bounded by Suez in the north, south to latitude 10° and east to longitude 78°. The area between 47°E

[10] Military forces, shipping associations, insurers and IMB have come together to produce the third version of *Best Management Practices*, which was released in June 2010. Worwood D., *A new anti-piracy bible,* Safety at Sea, 22, October 2010, vol. 44 no 500.

and 49°E remains the most dangerous for pirates attacks especially during the daylight[11].

4 HOW THE PIRATES OPERATE

The 2008 boom led to the fragmentation of piracy, and groups became smaller and more varied. There are groups of few former fishermen and a skiff or groups of about even 200 pirates involved in the business. In general, groups seem to be recruited from individuals with previous family or village ties. Sometimes a group has a tied family connections. There are also groups organized around a skilful leader and have no family ties at all. Each pirate group is usually a loose constellation around a pirate leader who is usually a veteran pirate, reinvesting funds in new pirate missions, who often functions as a fund raiser. The second way is a number of people coming together. In this case everyone brings his own food and guns, but the boat is owned by a specific person.

There are three basic modes of organization. The first is the whole operation is owned by one man who funds everything. In such cases, the owner agrees with the people involved in the mission on certain percentage of payment if a ship is captured.

The third way consists of a fund raiser who collects money from investors and then funds the pirate mission. In all three cases, the pirate leader should be well-connected and respected in the community, and thus able to draw upon his personal network for protection and problem solving[12].

How the mission is organized also influences how much it costs varying from multi-ship group, which usually is more expensive to organize into a small one-skiff group, which may need no more than $300 to run an attack. Smaller operations have less chances to be successful but on the other hand there are fewer people to share potential profits. Usually pirates retie when they collect $50,000 and more.

Surprisingly the technological resources available are limited. GPS systems and night vision goggles are used but not common. GPS and goggles are often ordered from local businessmen who travel to Dubai especially to buy them. Ship identification systems are very rarely used, sometimes pirates use so called spotters in ports to get the information about a vessel but the most common way is to observe the area and make an attempt to capture a spotted vessel (slow with low freeboard, preferably without passive security or barbed wire). The pirates tend to be self-financing and the money from hijackings is reinvested in new attacks. Additionally, former pirates that have invested their gains in legitimate business quite commonly reinvest in piracy[13].

An attack usually is conducted in a typical and following way: pirates spot flowing vessels with the latest technology. Choose the objective and board on large units, sufficiently fast to escape in the event of failure of the attack. The use of a pirate 'mother ship', carrying personnel equipment, supplies and smaller attack craft has enable attacks to be undertaken in a greater range from the shore. Somali pirates seek to place their skiffs alongside the ship being attacked to enable one or more armed pirates to climb onboard. Pirates frequently use long lightweight ladders to climb up the side of the vessel being attacked. Once onboard the pirates will generally make their way to the bridge to take control of the vessel. Once on the bridge they will demand that the ship slows down or stops to enable further pirates to board. If the crew does not let them onboard they start shooting. If the attacked crew agrees to let them in, they would probably spend several months in pirates' base waiting for overbought and ransom. If the crew does not let the pirates on board they fulfil their threats[14]. Attacks have taken place at any time of the day. However many attacks have taken place early in the morning at first light. They try to operate from and in areas, where local authorities have little or no power (central or south Somalia), because it allows them not to spend any money on bribes, which means more money to share. The money from the ransom, after paying all costs, is divided between the group that directly attacked the vessel and the group that guarded it afterwards. Hijackers get more than the guards. However, the highest share goes to the first person that boards the ship. The myths of piracy in the greater Gulf of Aden are many, but the average pirate group is a clan-based, low-tech group, consisting of former fishermen[15].

Statistics show that the aim of Somali pirates become all types of ships, their size is not able to discourage piracy. On the contrary, the larger vessels, the greater the risk, that they become the target of an attack. Hijacking of a large enterprise, carrying expensive goods entails huge profits for the snatchers, which obtain in exchange for the release of the crew and the ship. In 2008 in the Gulf of Aden and Somalia it could be noticed the largest, as yet, increase of tankers hijackings, carried out in a very large distance from the land, along the East coast of Africa. The purpose of each attack is to occupy a ship, but not every attack is successful one as mentioned be-

[11] Ibidem, 22.
[12] Hansen S. J., op.cit., 34-35.

[13] Ibidem, 36-37.
[14] *Best Management Practices to Deter Piracy off the Coast of Somalia and in the Arabian Sea Area,* International Maritime Organization, 9-10, London 2010.
[15] Hansen S. J., op.cit., 41.

fore. Every vessel with varying speed and low side becomes a potential attack target[16].

The report from 2008 with 111 incidents shows how dangerous area for navigation is the Gulf of Aden and the East coast of Somalia. The number of attacks in this area increased by approximately 250% compared with the year 2007 and 2005 (noticed accordingly, 44 and 45 attacks) and approximately five times in comparison with the year 2006 (only 20). The number of attacks in August 2008 only, at 19, and rising in November and October 15 and 16, respectively[17]. In 2009 there were 217 ships attacked with 47 vessels hijacked Somalia accounts for more than half of the 2009 figures. 2009 has however seen a significant shift in the area of attacks off Somalia. While the 2008 attacks were predominantly focused in the Gulf of Aden, 2009 has witnessed more vessels also being targeted along the east coast of Somalia[18]. According to IMB reports in the first half of 2010 there were 100 armed attacks reported off the coast of Somalia including 27 vessel hijackings[19]. The profits which the pirates make with the money are enormous for them, therefore, abandoning this activity in the country without prospects is practically only a wishful thinking.

5 PIRACY THE THREAT TO MARITIME TRANSPORTATION

Piracy in Somalia is a much greater threat than it might seem. The international community tackling this problem must bear in mind each potential hazard resulting from piracy. This threat can be classified in three aspects:

- the importance of piracy for international trade, and in particular the transport of oil;
- the danger for the environment;
- the potential terrorist threat.

Safety at sea has been seriously jeopardized in recent years, thanks to the emergence of incidents of piracy in key and strategic transit points, which undoubtedly the Gulf of Aden is. In addition to the direct impact on vessels, crew, cargo, as well as the maritime industry, piracy threatens worldwide commercial marine. Obviously, it is crucial that firms, which delays with the delivery of goods to the port of destination, will be losing money. Adding to this the costs paid in ransoms, piracy should be noticed as a serious threat to the trade from an economic point of view. The cost of freight rose from Rs4,000 ($132) to Rs5,600 ($ 185) for a 20ft container and Rs8,000 ($ 265) to Rs11,200 ($370) for a 40ft container. The line justified the tariff by citing a persistent risk of pirate attacks[20]. As it shows these consequences are not limited only to companies, whose vessels are hijacked but there are also serious concerns of the increase in costs of insurance premiums for vessels intending to go through the Gulf of Aden. Their growth, is not only caused by an ongoing risk of war, but also dramatically increasing number of hijacked units. During 2008, insurance premiums were raised ten times[21]. It is estimated that for the increasing number of passing vessels on that route the cost of insurance from the risk of war for 20,000 ships can reach even 109 400 million dollars. If the costs of an additional insurance become too burdensome for the company, or the transition by the Gulf of Aden seems too dangerous, the decision is taken about extension of the routes to Europe and North America, bypassing the Cape of Good Hope. Companies increasingly decide to bear higher costs associated with the time of arrival at the port of destination, fuel and crew, then risking hijacking vessels, endangering crew's life or paying higher insurance premiums. All the above, directly influence prices to cover the costs of transporting the goods. The Gulf of Aden is used to transport the oil from Arab countries and the already high price of this valuable in the XXI century raw material may be increased, due to the price of its transport and high insurance.

Large tankers passing through the Gulf of Aden, which are the most common pirates' goals, might pose a danger associated with spills of substances into the sea, so sensitive and important ecosystem. During the attack on the Japanese oil tanker *Takayama* (was targeted by pirates in April 2008), by 20 graph hole oil was leaking out into the sea. The consequences of this event could be much more serious, if not fast reaction to the leakage of the substance. It is necessary to keep in minds that the pirates and their actions are based on the basis of much better weapon. The use of firearms, including rocket-propelled grenades, in the direction of the vessel could cause a fire, run a vessel on the ground and even sinking, which in turn can induce ecological catastrophe destroying marine birds and animals for many years. The objective of the pirates is forced to pay the ransom note, if the crew puts the resistance, going to more radical methods, may lead not only to the death of innocent people, but also an ecological disaster.

---

[16] Trapla M., *Metody zwiększania bezpieczeństwa statku w kontekście wzrostu piractwa,* Master's thesis, Naval University, 44, Gdynia 2010.

[17] Reports on Acts of Piracy..., op.cit., 27.

[18] 2009 worldwide piracy figures surpass 400, http://www.icc-ccs.org/index.php?option=com_content&view=article&id=385:2009-worldwide-piracy-figures-surpass-400&catid=60:news&Itemid=51, 22.11.2010.

[19] *A new anti-piracy bible...,*op. cit., 22.

[20] *Pirate range ever more vast,* Safety at Sea, November 2010, vol. 44 no 501, 12.

[21] Costello M., Shipping insurance costs soars with piracy surge off Somalia, The Times, http://business.timesonline.co.uk/tol/business/industry_sectors/banking_and_finance/article4727372.ece/., 11.09.2008.

Piracy has become in recent years a very popular form of acquiring money and pirates have become 'heads' of international terrorism. It should be noted that currently there are no institutions relating to such events. Maritime terrorism must be treated seriously, and it was indicated by the attack on the American destroyer USS *Cole*, and killing 17 Americans. Creating a hypothetical situation in which terrorists would attack a VLCC tanker on the approach to the Suez Canal and would cause its sinking, the consequences of such incident could be multiplied. Ships waiting to pass the Canal would be queuing or heading for the Cape of Good Hope giving pirates even more possibilities of being attacked.

Terrorism at sea takes many forms:

- directed at military and civilian vessels[22];
- kidnapping, hostage-taking and boats, which are tender cards for terrorists;
- characterised by a high level of cruelty directed against the crew of a ship, and taken vessels become floating weapons.

A terrorist organization can allocate financial gains from piracy to sponsorship this type of activity around the world. It is suspected that Al-Shabaab, a terrorist group from Somalia gains money in this way[23]. The activities of the pirates in Somalia is becoming increasingly dangerous, and very often links with terrorists. The best solution is to prevent the worst scenarios than attempt to solve the problem after the escalation.

6 CONCLUSION

The problem with Somali piracy is one of the most important issue in the XXI century. Maritime transport has become the back bone of our economy and we cannot allow anybody or anything to hamper it or threaten, as we simply cannot afford it. Not taking any steps to fight or reduce this activity would mean that we have to pay extra money for longer routes, ransoms, costly equipment and expensive insurance. The world's economy has already been suffering problems since 2008 crisis, paying further costs may be very difficult for even well developed countries. Somalia is a very special country in terms of its internal condition, and as an international community we have to help the country to stand up on its feet or we leave it as it is and would pay even higher costs. The matter is complex and would take many different kinds of measures to stabilize the situation at sea but it seems not possible to tackle piracy without solving the problem inside the country. There are centres of power onshore in Somalia and they can be allies in the struggle against piracy; that is if they have power adjacent to the pirate bases and some interest in fighting it. Today, these centres of power are an untapped resource that could be used in this struggle. They could also be used to monitor pirate groups on shore, to register them, and to prevent piracy. However, there has to be something in it for the local partners, either through active fishery protection or through local purchases[24]. The question is if we –as the international community – are ready to pay the costs of this actions. If not we have to be prepared to pay for the actions undertaken, but not necessarily successful. It is up to us.

[22] As 6 October 2002, during the attack on MV Limburg Yemen ship attack was terrorism, BBC, http://news.bbc.co.uk/2/hi/middle_east/2324431.stm., 13.10.2003. as 6 October 2002, during the attack on MV Limburg

[23] *Terrorism Haven*, http://www.cfr.org/publication/9366/terrorism_havens.html , 12.09.2010.

REFERENCES

2009 worldwide piracy figures surpass 400, http://www.icc-ccs.org/index.php?option=com_content&view=article&id=385:2009-worldwide-piracy-figures-surpass-400&catid=60:news&Itemid=51, 22.11.2010.

A new anti-piracy bible, Safety at Sea, October 2010, vol. 44 no 500.

Best Management Practices to Deter Piracy off the Coast of Somalia and in the Arabian Sea Area, International Maritime Organization, London 2010.

Costello M., Shipping insurance costs soars with piracy surge off Somalia, The Times,

Hansen S. J., Piracy in the greater Gulf of Aden, Norwegian Institute for Urban and Regional Research, London 2009.

http://business.timesonline.co.uk/tol/business/industry_sectors/banking_and_finance/article4727372.ece/., 11.09.2008.

Pirate range ever more vast, Safety at Sea, November 2010, vol. 44 no 501.

Somalia country profile, http://news.bbc.co.uk/2/hi/africa/country_profiles/1072592.stm, 16.11.2010.

Somalia, http://en.wikipedia.org/wiki/Somalia, 16.11.2010.

Terrorism Haven, http://www.cfr.org/publication/9366/terrorism_havens.html , 12.09.2010.

The world factbook: Somalia, https://www.cia.gov/library/publications/the-world-factbook/geos/so.html, 22.11.2010

Trapla M., Metody zwiększania bezpieczeństwa statku w kontekście wzrostu piractwa, Master's thesis, Naval University, Gdynia 2010.

Wardin K., „Ocena zagrożeń bałtyckich strumieni transportowych działaniami terrorystycznymi", Belstudio Warszawa 2007.

Worwood D., Reports on Acts of Piracy and Armed Robbery against ships, Annual Report 2008, International Maritime Organization (2009) MSC.4/Circ.115

Yemen ship attack was terrorism, BBC, http://news.bbc.co.uk/2/hi/middle_east/2324431.stm., 13.10.2003.

Piracy hot spots, http://www.southchinasea.org/docs/Threats.pdf. 28.12.2008.

[24] Hansen S. J., op.cit., 62.

24. Preventive Actions and Safety Measures Directed Against Pirates in the Gulf of Aden Region

D. Duda & K. Wardin
Naval University, Gdynia

ABSTRACT: Piracy in the Gulf of Aden region became a real threat at the beginning of the 21st century for the safety of transport in the region. For these reasons, the international community have taken preventive actions and developed measures to be applied to fight piracy and increase safety in the region. These multilateral activities are based both on international efforts to improve the political situation in Somalia, where pirates have their bases, as well as the introduction of certain practices and procedures to combat piracy in the Gulf of Aden region.

1 INTRODUCTION

The global trade development caused, that within the last 20 years, on the oceans, began to appear more and more tankers, and other vessels that carry a bewildering variety of goods (starting from weapons up to food supplies). The world today transports almost 90% of its all freight by sea. On average, it is estimated that at any time of a day travels more than 10 million containers around the world. One of the biggest dangers that faces the contemporary world of shipping is piracy. The phenomenon concentrates, first and foremost, on waters bordering areas affected by lack of political authority or severe instability, bad economic situation and various problems of social nature. On the other hand another important factor, which significantly affects the appearance of the potential risk of the piracy, is the geographical location especially presence of gulfs, straits or archipelagos, which make piracy possible and very profitable business.

A perfect example of a state meeting all the above factors governing incurrence and the development of piracy is Somalia. This country for many years has been the place of many various armed conflicts, from small tribal clashes, by a series of conflicts with neighbouring countries, until the great civil war, which has been affecting the country since 1991. As a result, there has been a catastrophic destruction of infrastructure, agriculture and domestic industry. Through all these years, Somalia was struck many times by droughts and natural disasters, which have led to a lack of food and as a result of that to unimaginable famine. This difficult situation caused the break-up of the state structures in the flame of the war. Somali authorities asked for assistance the international community, which expressed its approval on several occasions by a number of the resolution. Finally peacekeepers were established in Somalia, with a goal to stabilized the tense situation in the country. All of these missions failed to help the situation and the only advantage of them was to show how difficult are the realities in this part of the world.

The absence of alternative sources to acquire the means to live stimulates the Somali population to take up piracy business and related activities. In that reality piracy seems the only possibility of survival. In order to prevent acts of piracy and armed robbery it has been decided to regulate legally, what piracy and armed assault is and then what the difference between these two offences is. The list of offences has been introduced for this purpose as well as criminal sanctions for committed acts. There were also introduced the principles of extradition, legal ways of accusing presumed criminals and how to pass the international cooperation. In addition, the United Nations (UN) Security Council and the European Parliament in connection with the deteriorating of political situation and increase in attacks in Somalia, have decided to post the relevant resolutions containing provisions concerning improvement of safety and inform about the current situation in this area.

The problem of modern piracy, particularly in the area of the Horn of Africa, through political implications is extremely difficult, and therefore, precise and comprehensive solution is not easy to work out. It is related to the long-term process of decision-making in order to prevent criminal acts in this area. Requires the cooperation of both local and international centres, which must overcome a number of cultural, political, economic barriers and social conditions. It is also very important to act instantly and try to keep maritime transport as safe as possible. For these reasons some measures have been undertaken immediately.

2 ORGANISATION OF MARITIME TRANSPORT IN A SECURE AND SET CORRIDOR IN THE GULF OF ADEN

During the early months of 2008, security in the Gulf of Aden was almost solely provided by the Combined Maritime Forces ("CMF") Combined Task Force 150 ("CTF-150"). At various times CTF-150 has comprised vessels of the United States, the United Kingdom, Canada, France, Germany, Australia, Italy, The Netherlands, New Zealand, Spain, Portugal, Denmark, Pakistan and other nations. But piracy has not been its main target—which is general maritime security as part of the War on Terror—and CTF-150 is thinly spread over not just the Gulf of Aden but also the Gulf of Oman, the Arabian Sea, the Red Sea and a large part of the Indian Ocean; a total of 2½ million square miles. To cover this vast area CTF-150 usually has about 14 ships, including supply vessels[1].

During 2008 the number of pirate attacks in the Gulf of Aden dramatically increased, especially along the coast of Yemen. In August 2008, as a result of pressure from the International Maritime Organization and other bodies, CTF-150 established a Maritime Security Patrol Area (MSPA) in the Gulf of Aden (Figure 1.), with the intention of channelling merchant vessels through a corridor that would in theory afford greater safety, because defensive measures would be more effective when concentrated in a smaller area. Thanks to the coalition patrols of warships and aircrafts, the safety of all vessels travelling in the basin was ensured. Easier control, monitoring, and the possibility of faster response, scored big chances of international forces in the fight against widespread piracy and everything seemed to coming back to normal. But all these measures, although ensuring the safe passage of some merchant ships, and preventing some boardings, could not avert a further large number of hijackings and an even greater number of unsuccessful attacks. The situation began to worsen and pirates started to attack more often and farther than ever after hijacking of MV *Sirius Star* - an oil tanker – in November 2008[2]. It was clear that the MSPA did not achieve the expected results, and the passage of the vessel inside the corridor involved increasing risk. There were not enough warships patrolling MSPA, and it was quite a widespread area.

Figure 1, Maritime Security Patrol Area 2008 (MSPA), source: http://www.eaglespeak.us/2008/08/gulf-of-aden-martime-security-zone.html, 20.11.2010.

In connection with the escalation of action in the Gulf of pirate, at the beginning of 2009, **International Recommended Transit Corridor (IRTC)** was established (Figure 2.), and boats were advised to travel in convoys.

The creation of a new corridor was to group ships, in order to give them additional protection not only from the army, but also from each vessel included in the group. Mutual observation of vessels, was giving additional protection, so important during the transition. The greater number of units patrolling the corridor, was to discourage Somali pirates to attack ships.

IRTC has to secure safe passage through the Bay. The zone is plotted by the centre of the Gulf Aden and ships move one rate for the entire length of the corridor. In addition, the risk of attack has been limited by the separation of the corridor into two parts, for traffic on the East, and for traffic on the West, by separating the two tracks of two mile safety zone. Entry in the corridor is strictly controlled, entering the boundary corridor for specific times, depending on the speed of the vessel. Delimitation of the corridors for the movement to the East and West, as well as narrowing them, and the assigning of the vessels according to their speed possibilities to the time of entry into the zone, greatly helped forces of military coalition in this area to make the area safer to pass[3].

[1] Knott J., *Somalia, The Gulf of Aden, And Piracy*, http://www.mondaq.com/article.asp?articleid=72910&print=1, 21.11.2010.

[2] Grinter M., *Piracy: what is the solution,* 15-16, Maritime Asia December 2008/January 2009.

[3] Trapla M., *Metody zwiększania bezpieczeństwa statku w kontekście wzrostu piractwa,* Master's thesis, Naval University, 57, Gdynia 2010.

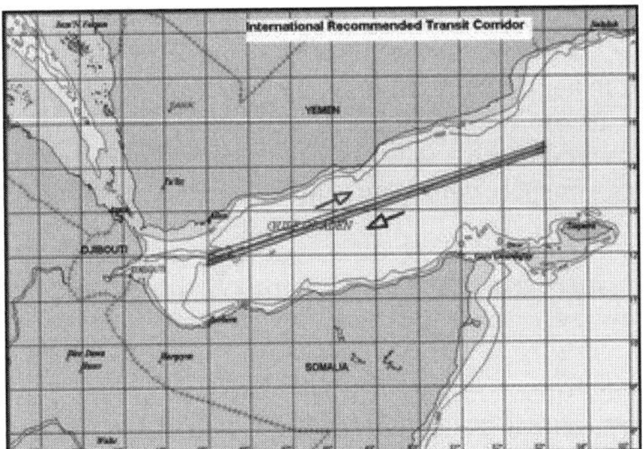

Figure 2, International Recommended Transit Corridor, source: http://asianyachting.com/news/PirateCorridor.htm, 20.11.2010.

Since the introduction in 2009 of the Internationally Recommended Transit Corridor for all ships passing thorough the Gulf of Aden, cruise-goers can feel much more secure against the threat of Somali piracy. Potential cruise passengers should also be reassured about such matters as travel insurance, changes to the cruise excursion schedule and the unobtrusive nature of the protection offered.

To make sure that situation is properly handled in January 2009 was created the CMF's Combined Task Force 151 ("CTF-151"), commanded by US Navy Rear Admiral Terence McKnight, with a specific mandate to counter piracy operations in and around the Gulf of Aden, the Arabian Sea, the Indian Ocean and the Red Sea, thereby releasing CTF-150 to carry out its original task of anti-drug, anti-smuggling, and other general maritime security operations[4].

3 OPERATION ATALANTA

Although CTF-151 helps efficiently to keep the Somali waters and the Gulf of Aden secure, it is not the only military mission in the area. Since 8 December 2008 the European Union (EU) has been conducting a military operation to help deter, prevent and repress acts of piracy and armed robbery off the coast of Somalia. This military operation, named EU NAVFOR Somalia-operation ATALANTA, was launched in support of Resolutions 1814, 1816, 1838 and 1846 which were adopted in 2008 by the United Nations Security Council. Its aim is to contribute to:

- the protection of vessels of the World Food Programme (WFP) delivering food aid to displaced persons in Somalia;
- the protection of vulnerable vessels sailing in the Gulf of Aden and off the Somali coast and the deterrence, prevention and repression of acts of piracy and armed robbery off the Somali coast.

[4] Knott J., op. cit.

This operation - the European Union's first ever naval operation - is being conducted in the framework of the European Security and Defence Policy (ESDP)[5].

Operation ATALANTA's mission is to:

- provide protection for vessels chartered by the WFP;
- provide protection for merchant vessels;
- employ the necessary measures, including the use of force, to deter, prevent and intervene in order to bring to an end acts of piracy and armed robbery which may be committed in the areas where they are present.

The EU Operational Headquarters is located at Northwood, United Kingdom. The Political and Security Committee exercises the political control and strategic direction of the EU military operation, under the responsibility of the Council.

The operation was initially scheduled for a period of twelve months. During that period more than twenty vessels and aircraft took part in EU NAVFOR. At present it has been extended by the Council of the European Union until December 2010, and again for another two years, until December 2012.

The EU NAVFOR Operation Atalanta consists of units from Belgium, France, Germany, Greece, Italy, Luxemburg, Netherlands, Spain and Sweden. Contributions from third countries such as Norway, are participating as well. Malta, Portugal and United Kingdom are also participating. There is also a close cooperation with Russian, Indian, Japanese and Chinese vessels[6].

The composition of EU NAVFOR changes constantly due to the frequent rotation of units and varies according to the Monsoon seasons in the Indian Ocean. However, it typically comprises 5 - 10 Surface Combatants (Frigates/Destroyers), 1 Auxiliary and 3 Maritime Patrol Aircraft. Units are drawn from the contributing nations. The Force Head Quarters vessel rotates on a four monthly basis.

The operation can arrest, detain and transfer persons who have committed, or are suspected of having committed, acts of piracy or armed robbery in the areas where it is present and it can seize the vessels of the pirates or armed robbers or the vessels caught following an act of piracy or an armed robbery and which are in the hands of the pirates, as well as the goods on board. The suspects can be prosecuted either by an EU Member State or under the EU-Kenya agreement, which gives the Kenyan authorities the right to prosecute[7]. A website is used

[5] EU naval operation against piracy (EU NAVFOR Somalia - Operation ATALANTA), http://www.consilium.europa.eu/uedocs/cmsUpload/090313FactsheetEU-NAVFORSomalia-v3_EN.pdf, 25.11.2010.

[6] European Union Naval Force Somalia - Operation Atalanta http://www.eunavfor.eu/about-us/, 25.12.2010.

[7] Ibidem.

in coordinating both merchant shipping and military activity. This approach has been welcomed by the merchant shipping industry. Merchant vessels that follow EU NAVFOR recommendations run a much smaller risk of being attacked and/or captured.

Since the start of the operation the number of attacks by pirates has greatly diminished. This is linked to the dissuasive presence of the vessels of the EU NAVFOR ATALANTA force and to the self-protection measures which have been put in place for merchant shipping at the recommendation of the European naval force.

Operation EU NAVFOR is part of the global action conducted by the EU in the Horn of Africa to deal with the Somali crisis, which has political, security and humanitarian aspects.

The EU supports the Djibouti process for peace and reconciliation in Somalia, facilitated by the UN. As the effectiveness of protection measures employed within the Gulf of Aden has increased, pirates have started to operate in previously unused areas to avoid interdiction by EU NAVFOR and other counter-piracy forces. Through the 2009 inter-monsoon periods it became evident that pirate action groups were operating at ever greater range to avoid detection. In light of these changes, EU NAVFOR has increased its area of operations to maintain pressure on the pirates and to continue to constrain their freedom of action. In doing so, EU NAVFOR endeavours to ensure that legitimate maritime traffic within the region continues to receive the best protection possible. This procedural change allows EU NAVFOR units to operate more effectively further east in the Indian Ocean, giving them a greater ability to disrupt and deter pirates in this vast area[8].

It is very difficult to evaluate the operation which has still been in progress. The basic idea behind this strategy is to deter pirates by making it harder to hijack ships. The strategy is mainly an offshore focussed strategy, with little emphasis on onshore measures to prevent piracy, however, in practice it seems to be combined with an onshore, centralized state-building strategy. It is very difficult however to build a stable central government as its power rage is limited to several city quarters in the capital city Mogadishu. Without employing local institutions, authorities fighting piracy might be a very difficult, expensive and long process, and nobody can guarantee its success.

The containment approach dominates the approaches to Somali piracy today; major funds are being used on it in order to contain piracy. The European Union for example uses "The European Union's Joint Strategy Paper for Somalia" and pledged €212 million for development assistance from 2008-2013, while the EU's joint naval endeavour, Operation ATALANTA, planned to spend an estimated $450 million in one year only[9]. The result so far can be described as satisfactory but far from very good as we still can hear about attempts of attacks or successful attacks in the area.

4 DIRECT MEASURES TO PROTECT SHIPS FROM PIRACY

In the situation of the still growing acts of piracy, of the coasts of Somalia mainly in Gulf Aden, the International Maritime Organisation (IMO) has decided to issue a series of recommendations on good practices in the field of the fight against piracy. Ship-owners and masters of vessels in the area are recommended to sign their ships on the website of the Centre of maritime safety on the Horn of Africa-MSC (HOA) before the vessel enters the Gulf Aden. It gives the possibility of continuous monitoring by EUNAVFOR – ATALANTA, and obtaining all necessary information about the situation in the Gulf of Aden. Unfortunately as IMO delivers every third ship does not follow this routine and endangers itself and the crew.

The European Commission calls on EU Member States, to use all available means of protection, which are aimed at combating hazards while crossing the Gulf of Aden. At the same time calls on the EU States, to disseminate, control and up-date best practices, aimed at combating acts of piracy and armed attacks in the Gulf of Aden.

IMO issued a series of recommendations aimed at avoiding slowdown, and inhibit acts of piracy and armed assault along the coast of Somalia and the Gulf Aden, intended for companies, operators, captains and crew members. The practices were accepted by international members of maritime sector (eg. INTERTANKO - International Association of Independent Tanker Owners, ICO - International Chamber of Shipping, BIMCO – The Baltic and International Maritime Council, IMO – International Maritime Organization and many others)[10].

Generally speaking it is being advised to use the IRTC and to enter the gulf in groups at scheduled time which gives the possibility to international forces to escort the vessels. They should avoid entering territorial water of Yemen as they are not protected by international forces. Apart from that the ships should transit the waters at night as the number of attacks is much lower at that time.

At the time of planning the transition through the high risk area, the company together with the captain of the ship must make the relative risk assessment,

---

[8] EU NAVFOR, Press release, http://www.eunavfor.eu/2010/09/eunavfor-atalanta-area-of-operation-extended/, 22.09.2010.

[9] Gilpin, R., "Counting the cost of Somali Piracy" *United States Institute of Peace Working Paper* April 2009.

[10] Trapla M., op. cit., 78-79.

taking into account all the most up-to-date information from the area of the threat. This assessment will allow to estimate the likelihood of attack by pirates, and take additional outside regulation measures in favour of the fight against piracy. Owners are obliged to register their ships on the website of the MSC (HOA) – www.mschoa.eu, since it is the source of many necessary and most up-to-date information needed when planning a journey in waters of Somalia and the Gulf of Aden. The captain is obliged to ensure that all procedures associated with the passage of a vessel by a vulnerable area were made. He must make sure that the company reported to MARLO – Maritime Communications Office and UKMTO Dubai - the United Kingdom Maritime Trade Operations intention of entry of the vessel in international Transit Corridor – the IRTC, five days before the planned entry. If such notification has not been sent, the captain is required to complete this obligation. The captain must accurately inform the crew of the planned passage and anti-piracy protection measures. The captain decides also if the AIS (Automatic identification System) should be turned on or not. A very important element in secure against pirate attack is blocking access to the ship. In the first instance you should protect and control access to the bridge of the captain, the bridge, the engine room, and all cabins inside the vessel.

It is recommended that there is a special dedicated room onboard -a citadel 'a hardened' secure space in the ship, which fulfils the role of the assembly at the time when the risk of attack has been pirated. It is designed to block and delay access to the control of the ship by pirates. Although in 2010 military forces have freed three vessels by conducting 'opposed boardings' while the crew was safely sealed in citadels. Despite this, shipping operators are warned that they should not treat a citadel as a panacea and that a proper risk analysis is essential[11]. Requirements for the placing of the citadel and how to use it are the newsgroups, for this reason, masters of vessels, on a regular basis should verify this information via the MSC (HOA).

Another a very popular way of thwarting a pirates attack is the use of water and fire hoses and hose to clean the deck. It is recommended that they are distributed over the whole length of the deck ready for use, and the pump, supplying water to the hydrants should be in on, while passing through high risk areas. High pressure water stream addressed in the attackers, effectively can refute the attack.

Additional protection for the ship may be applicable Razor Ship. This is a system of cables wired along the side of the ship and low placed to enter at the stern. This is one of the essential elements of the passive defence before the attack, which can be used while passing the sea, at the time of berthing at anchor and in port. The system has many advantages, thanks to which it is so popular on ships, and they are:

- low cost;
- strong psychological barrier to overcome, not ease to overcame for the attackers;
- easy assembly and dismantling (such a security system is only recommended in risk areas);
- high efficiency and the possibility of prolonged use; practical packaging allows the quick and safe positioning of the crew to wire the whole length of the ship, and facilitates storage in marine conditions .
- system of passive protection (approaching at your own risk);
- easy to link individual lengths of wire (in the case of securing large surface)
- a good tool for protection against pirates, together with the LRAD (Long Range Acoustic Device - an acoustic sound waves device which produces, long range acoustic wave energy of 151dB, for comparison, the threshold of pain for a man is about 120dB, it is applied in immediate distress), it is one of the more effective systems to fight[12].

Another very helpful device in defence against piracy are Counter Piracy Net. This is a substantial change in the field of passive defence. Plastic containers with the network should be on the perimeter of the entire vessel, in the proper distances. In the event of danger, the security shall be released and to discharge the net on the surface of the water or just below it, by pulling along the freeboard and stern up to 50 metres. In addition, the networks are equipped with a orange buoys, to act as a deterrent and visually warning of danger.

BEA Systems is another way to direct the fight against piracy. BEA Systems has pursued its expertise from the scope of the defence, safety at sea and airspace sector to develop a technology that enables the detection of small units and identify actions of pirates in a radius of 25 km, such distance is sufficient for this that the crew can prepare themselves to take appropriate action, in order to avoid an attack

BAE System includes such devices as:

- high frequency radar to detect small boats in a radius of 25 km;
- wide area surveillance system, the system makes it easy to detect traffic unit and has alarms levels of risk;
- passive radar identification system (PRISM), intended to provide early warning against unidentified ships;
- equipped with better lighting, giving the ability to detect and deter intruders who are in the vicinity of the ship during the night[13].

[11] Worwood D., *A new anti-piracy bible,* Safety at Sea, 23, October 2010, vol. 44 no 500.

[12] Trapla M., op. cit., 85-86.

[13] Ibidem, 87-89.

Another effective weapon in the fight is an ion cannon. It supposed to send ion beam with such force that any electronic systems, responsible for work of the weapon systems or propulsion and steering were unable to use. Before the struck boat comes back to normal, the vessel that became the target of an attack gets the time to escape[14].

As we can see there are many different, passive possibilities to protect the ship from pirate attacks. All of them are probably quite efficient but they must be put in practice otherwise they are just useless gadgets. This is directly connected with money spend on equipment of modern ships, which should be able to have such devices installed if necessary. The money spent on passive protection, training programmes and other things connected with better security is a well spent money, which may save the crew. It should be noticed that the situation in the Gulf of Aden and Somali Basin has become very difficult in such a short time that many of ship-owners, captains and crew members do not want to accept as true that passing in the described area is dangerous in spite of the presence of international forces from CTF 151 and EU NAVFOR.

5 ACTIVITIES UNDERTAKEN OUTSIDE PIRACY ZONE

IMO specifies how the crew should behave while passing through the area of the Gulf Aden and how should the vessel be carried out. Primarily, the master should avoid areas in which the crew is exposed to the risk. Presence on the external parts of the deck should be limited to the minimum from dusk to dawn, however, be aware of the continuous performance of watch keeping.

Vessels should be addressed through the Centre of maritime safety and the licence of the Horn of Africa-MSC (HOA) to the corridor liner - the IRTC, at appropriate times, depending on the speed of the vessel. The captain should be in constant contact with the above mentioned centre and keep track of the information provided by the institution on the website.

If the route of the ship is not passing through the IRTC, this unit should move with full engine, not slower than 18 knots. Transition by the Bay as far as possible should be carried out after dark.

It is important that both vessels entering and outgoing from the IRTC do not switch off navigation lights, while they limit lights on-board. Additional lighting of a ship using a remotely operated head-lamps is consistent with the law, but they may not be jeopardized in any way the security of the maritime traffic.

During the voyage, special attention should be given to small boats, which are frequently used to attack by Somali pirates. Their photographs and description should be on board the vessel. When the crew enter into suspicion about the approach of this type of craft to the vessel, the risk assessment should be made immediately and a report of this event should be sent to UKMTO Dubai and IMB PRC using the after the attack report. The pirates boat should be immediately informed that they had been noticed using lamps, alarms, and the corresponding movements of the crew. If possible use of all available non-lethal types of weapons, assess their advantages and usefulness in a given situation, and the exposure of the ship concerned.

It should be mentioned that pirates develop very quickly and adjust their actions to changing conditions which can be seen for instance in stretching the area of attacks. This forces International Maritime Bureau (IMB), military forces, shipping associations and insurers to take suitable actions by producing the third version of Best Management Practices (BMP3).

Released in June 2010 , the new, pocket-sized BMP3 contains everything that ship-owners, operators and masters need to know about deterring attacks[15]. EU NAVFOR commented the BMP3 to be a very helpful tool to avoid attacks especially because in multicultural nature of crews and the fact that standards can vary, it is important to have one common way of action in case of pirates attacks. The booklet is distributed free of charge to ships.

It contains additional advice on ship-protection measures- aimed at counteracting the latest pirate tactics. It also includes a copy form of the UKMTO form for vessel position reporting.

As mentioned above reflecting the increasing range of pirate attacks, the high-risk area defined in BMP3 has been expanded beyond the Gulf of Aden to the area bounded by Suez in the north, south to latitude 10° and east to longitude 78°. The area between 47°E and 49°E remains the most dangerous for pirates attacks especially during the daylight.

All the precautions have not entirely prevented vessels being attacked in or near the transit corridor. It should be barred in minds that the planning, constant alertness is crucial for the safety of any ship travelling at the described waters.

If pirates were able to get on board it is very important that any action taken by both the master and the crew of the vessel were directed to:
- ensure the safety of all persons on board;
- maintain control over the vessel;
- leaving the ship by attackers.

[14] Source: *Ion cannon* http://www.sluisvan.net/uzbrojenie_statkow,energetyczne, 20.11.2010.

[15] Worwood D., op. cit., 22.

In case of both, the risk of attack, and in the case of invading pirates on board a vessel, this fact should be immediately reported to UKMTO Dubai, and if possible to the ship-owner. This should be done before the introduction of the attackers on the bridge. It is very important that the whole crew is together (excluding personnel on the bridge) and remained in the same location on the ship, They should show no resistance and do not behave in a daring way. If there is a citadel, it should be equipped with necessary measures and use them in the event of intrusion attempts of the pirates. In addition, the crew should keep away from any illuminators, manholes, and does not try to stop the pirates from boarding the ship. Be aware of the emergency communications available from inside the citadel and as soon as possible connect with the relevant.

It is extremely important to remember these rules, because they may save crews' lives in a critical situation. It would be very helpful to bear in mind such points:

- ship operators should register at www.mschoa.org and submit a vessel movement registration-form;
- sent a UKMTO vessel position reporting form (included in BMP3);
- report transit details regularly;
- define the ship's AIS policy;
- keep emergency contact numbers near;
- read BMP3 implement protection measures and test anti-piracy procedures before the high-risk area;
- maintain crew vigilance and awareness, avoid complacency;
- drill crews in what to expect and how to react;
- always use the IRTC and Gulf of Aden group transit;
- if using a citadel, ensure that all crew members are safely inside;
- if under attack keep the ship moving.

CONCLUSION

No matter what temporary precautions we are going to take and how alert we stay in dangerous waters, it is important to remember that fighting piracy must take place not only offshore only, leaving escaping pirates to shelter and ride off the storm to try another day. To fight them we must use many methods and try to engage local institutions. It is true that in Somalia there is lack of powerful central government, but local institutions function pretty well and they should be used in this struggle. So far, focus on a centralized solution has limited the international fleets' access to information from onshore sources. It has also limited the international fleets' ability to cooperate with entities that de-facto hold power close to the pirate bases[16]. There is no single solution to Somali piracy, and none of the described approaches is entirely successful nor without merit. On the other hand we have to be aware that European-founded operations to combat Somalia-based piracy could be hit by major public spending cuts[17]. Many countries in Europe have serious problems with constructing their budgets and extra money spent on piracy might be a difficult overweigh impossible to bare for next year or years to come. It can also be heard that shipping must stop relaying on a limited military presence and deal with the problem through proper crew training and vessel design[18]. It is of course a solution but there are also voices saying (especially BMP3) that the use of special armed guards on board the ships is not the best solution.

Total elimination of piracy through the constant control of the Gulf of Aden by using the ships involved in the ATALANTA operation and others is rather unlikely but its limitation should contribute to resolve these issues and gain stability. However, this cannot be the only step done by the 'West' in order to ensure safety in this part of the world. We have to offer Somali pirates some alternatives, work opportunities otherwise ex-pirates are likely to slip back into a life of maritime crime.

REFERENCES

Cuts and counter-piracy, Safety at Sea, November 2010, vol. 44 no 501.

Davis N., Take responsibility for stopping piracy, Lloyd's List, October 2010, No. 60.273.

EU naval operation against piracy (EU NAVFOR Somalia - Operation ATALANTA), http://www.consilium.europa.eu/uedocs/cmsUpload/090313FactsheetEU-NAVFORSomalia-v3_EN.pdf, 25.11.2010.

EU NAVFOR, Press release, http://www.eunavfor.eu/2010/09/eunavfor-atalanta-area-of-operation-extended/, 22.09.2010.

Gilpin, R., "Counting the cost of Somali Piracy" United States Institute of Peace Working Paper April 2009.

Grinter M., Piracy: what is the solution, Maritime Asia December 2008/January 2009.

Hansen S. J., Piracy in the greater Gulf of Aden, Norwegian Institute for Urban and Regional Research, London 2009.

http://www.mondaq.com/article.asp?articleid=72910&print=1, 21.11.2010.

International Recommended Transit Corridor http://asianyachting.com/news/PirateCorridor.htm, 20.11.2010.

Ion cannon, http://www.sluisvan.net/uzbrojenie_statkow,energetyczne, 20.11.2010.

Knott J., Somalia, The Gulf of Aden, And Piracy,

Maritime Security Patrol Area 2008, http://www.eaglespeak.us/2008/08/gulf-of-aden-martime-security-zone.html, 20.11.2010.

[16] Hansen S. J., Piracy in the greater Gulf of Aden, Norwegian Institute for Urban and Regional Research, 3, London 2009.

[17] *Cuts and counter-piracy,* Safety at Sea,13, November 2010, vol. 44 no 501.

[18] Davis N., *Take responsibility for stopping piracy,* Lloyd's List, 6, October 2010, No. 60. 273.

Trapla M., Metody zwiększania bezpieczeństwa statku w kontekście wzrostu piractwa, Master's thesis, Naval University, Gdynia 2010.

Worwood D., A new anti-piracy bible, Safety at Sea, October 2010, vol. 44 no 500.

European Union Naval Force Somalia - Operation Atalanta http://www.eunavfor.eu/about-us/, 25.12.2010.

25. Technological Advances and Efforts to Reduce Piracy

M. Perkovic, E. Twrdy, R. Harsch & P. Vidmar
University of Ljubljana, Faculty of Maritime studies and Transport, Slovenia

M. Gucma
Maritime University of Szczecin, Poland

ABSTRACT: The technological contributions to the reduction of piracy not only involve implementations of recent technological advances, but, importantly, the dissemination of the education required to apply current and future technologies, particularly in those states in the regions where piracy is rampant. To this end, the EU's MARSIC project, with the stated aim of enhancing security and safety in the Gulf of Aden and the western Indian Ocean through '..information sharing and capacity building, (and) highlighting regional cooperation' (Marsic 1st monitoring report, 2010) has recently been inaugurated. The Faculty of Maritime Studies and Transport of the University of Ljubljana, and the Maritime University of Szczecin, as partners in this project, will bring to bear both the most advanced technological applications to maritime affairs of satellite imagery, simulation, and risk assessment, and guarantee their utility through the transfer of knowledge. In Yemen and Djibouti, maritime stations will be established, personnel trained, and a sustainable level of expertise eventually left in place. Interest in such projects has also been expressed by maritime experts in Tanzania and Kenya. The advantage this approach has over other donor-supported solutions begins with regional involvement and an inclusive approach, its ultimate success to a large degree dependant on factors external to the project such as financial incentives for the nations of the region to protect European and Far East Asian shipping. The project is closely coordinated with a parallel EU-funded project executed by European Commission's Joint Research Centre (JRC) on maritime surveillance technologies application in the region.

1 INTRODUCTION

One general difference between the social sciences and the physical sciences is that the physical sciences have control of their laboratory and in fact have to determine their own variables. The social sciences are confronted by a plethora of variables that they artificially order for the purposes of study. The current confrontation with piracy poses a challenge for both classes of science. As piracy is endemic in the sea within range of Somalia, the main challenge for the social scientist is in determining how to convert Somalia from its present virtually stateless condition to a viable economic entity that values international law and provides an economic environment in which piracy is no longer a sensible option for its citizens. For the physical sciences, particular those acting at the behest of stakeholders in the global shipping economy, it is a great challenge to approach the problem without considering the challenges typically restricted to the social sciences. In fact, the case might even be made that the scientific efforts described in this paper place the physical scientist in the position of acting as an ancillary to global policing activities. This, of course, seems to diminish the activities of the scientist; however, it is only another way of saying that science applied in real time to real problems necessarily involves the recognition of the daunting task of striving to create real methods that can be implemented as real solutions at virtually the same time. As an arm of the major economic interests, the scientists who operate in the arena of piracy are disposed to devise methods to counter the threat piracy poses to global shipping, which primarily affects western European, North American, and East Asian economies.

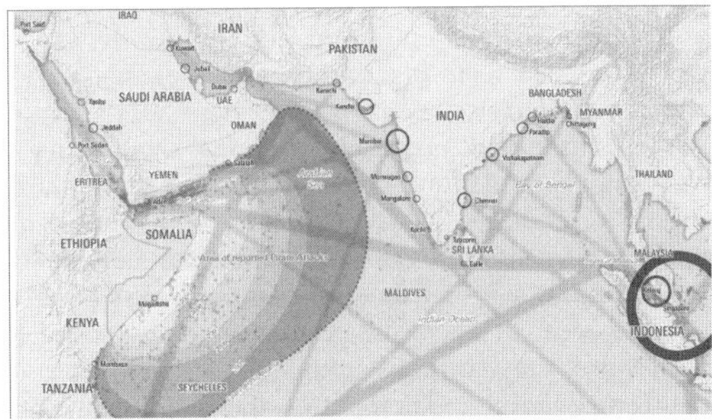

Figure 1: Extent of 2010 Pirate attacks (*Threat Map, 2005-2010*)

The importance of this consideration will be highlighted shortly. First, it should be of benefit to anyone concerned with piracy in the vicinity of the Red Sea, eastern Africa, and the Arabian Sea, to be aware of the point of view of the peoples bordering these waters. The global economic shift that took place once Europeans decisively established themselves as the powers of the Indian Ocean is not viewed passively as an inevitable manifestation of a superior economic system by the people of India, Iran, the Hadramaut, Somalia, Egypt, and so on (*Chaudhuri 1985, Das Gupta 2001, Floor 2006, Hall 1998, Hourani 1995, Braudel, 1995*). In the west we are taught that Vasco da Gama was the first to round the cape of Africa. We are not taught that, in fact, he was the harbinger of massive forced change, disruption to a trade extant for multiple millennia, and eventually the decimation of economies that have yet to recover from the onslaught. Nor are we taught that da Gama himself returned to India a second time and committed atrocities of unimaginable brutality without provocation (*Hall 1998, Harsch 2011*). This is not to say that memories of such depredations linger at the forefront of the minds of those in this region, it is fair to say that this history provides a measure of the context of their thought development.

Focusing specifically on Somalia and Somalians, as the Somalians are the greatest threat to global shipping (among pirates), again the physical scientist would do well to consider some social circumstances. First, as Somalia is without a coast guard or navy of any effect, their extensive waters are subject to extensive pirate fishing. This problem is not limited to Somalia, but extends to all troubled developing coastal countries. According to a report by the Environmental Justice Foundation, "Illegal, unreported and unregulated (IUU) fishing is one of the most serious threats to the future of world fisheries. It is now occurring in virtually all fishing grounds from shallow coastal waters to deep oceans. It is believed to account for a significant proportion of the global catch and to be costing developing countries up to $15bn a year." (*Guardian, 2009*). Yet more nefarious is the illegal dumping of hazardous wastes in waters of vulnerable countries: "In 1991, the government of Somalia collapsed. Its nine million people have been teetering on starvation ever since – and the ugliest forces in the Western world have seen this as a great opportunity to steal the country's food supply and dump our nuclear waste in their seas. Yes: nuclear waste. As soon as the government was gone, mysterious European ships started appearing off the coast of Somalia, dumping vast barrels into the ocean. The coastal population began to sicken. At first they suffered strange rashes, nausea and malformed babies. Then, after the 2005 tsunami, hundreds of the dumped and leaking barrels washed up on shore. People began to suffer from radiation sickness, and more than 300 died" (*Hari, 2009*). The point is not to shift the focus from piracy and the need to find methods to reduce its impact, rather to ensure that no one involved in efforts to combat the scourge proceed unaware of a strikingly important aspect of its context. Practically, this means that involvement in such European Commission projects as MARSIC (*Gaullier, 2010*), which plans to open functioning information sharing centers in Djibouti and Yemen, providing advanced training and equipment to these centers designed to be fully operated by local professionals, are aware that, for example, consideration is given to the need to enhance Yemeni and Djibouti economic prospects through this effort, to make this perhaps the first step in including especially Yemen, the poorest of Arab countries, in the global maritime trade complex in such a way that its benefit extend far beyond the fortuitous circumstance of having a port, Aden, in a strategic location.

2 MARSIC

The EU MARSIC program is designed as a step toward improving the safety and security of shipping in the western Indian Ocean and the Gulf of Aden, areas of intensive pirate activities, through the creation of Regional Maritime Institutes (RMI) in the region, including Djibouti, Yemen, Kenya, Tanzania, and other signatory countries of the Djibouti Code of Conduct, which pledges these countries to combat piracy through information sharing on the regional level. MARSIC intends to offer the resources and training to provide these RMIs the latest technology applicable to vessel traffic monitoring, information sharing, and situational analysis based on up-to-date data bases. Merely implementing AIS would be an advancement in most countries, but the challenge addressed by the authors of this paper also include applications of satellite imagery, risk assessment, and the upgrading of legal systems.

Notwithstanding the background of the explosion of piracy in the target areas and the legitimate grievances of peoples of the region, efforts to reduce the instances of piracy and the economic effects of these instances are urgently necessary. One conclusion arrived at both by IMO and the regional littoral countries of the infested areas is that regional involvement is necessary. The MARSIC program intends to support regional efforts in part through the sharing of the most advanced technological aids available. At both the EC-JRC and the University of Ljubljana Faculty of Maritime Studies and Transport the uses of satellite imagery for maritime purposes has been an ongoing project. Advances have been rapid and effective. For instance, it is now, largely through the efforts of scientists at both these institutes, possible to locate oil spills and backtrack towards the identification of illegal polluters (*Perkovic, et al. 2010*), as well as to determine causes and effects that would otherwise be mystifying (*Ferraro, et al. 2010*). One of the frustrating aspects of the constant increase in instances of piracy is that the traditional guardians of commerce at sea, navies, have been unable to reduce the number of instances, even though some 25 ships, assisted by many patrol aircraft and even submarines from about 20 different countries patrol the seas. Satellite technology applications, risk analysis, maritime law, VTS, and maritime communications techniques will all be part of the effort to create effective RMIs in the region.

3 MARITIME DOMAIN AWARENESS

To combat piracy, or in fact any illegal or undesired activity at sea, it is in the first place necessary to be aware of what is going on, and be alerted to threats and problems. Currently, the countries in the Horn of Africa region are either lacking such awareness, financially incapable of contributing to a resolution of the problem, or politically disinclined to do so. This is understandable, as up to now there was no pressing need to invest in an (expensive) maritime surveillance and patrol infrastructure. Now, however, the situation has changed, with, as explained above, not only piracy, but also illegal fishing, waste dumping and trafficking affecting the regional security situation. At present, foreign navies fulfill the task of providing maritime security (within the stringent bounds of international law) and the maritime awareness needed to do so, using the top-level technical means at the disposal of the military. In the long term, however, such a foreign presence is not sustainable or desirable. The challenge is to find ways, technical means, for the national authorities in the region to use in building up their maritime domain awareness by themselves. The Pilot Project on Piracy, Maritime Awareness and Risks, carried out by the JRC, explores this problem (*Greidanus,2011*). The project considers a wide variety of sources of information on ship traffic, ranging from cooperative (reporting) systems (LRIT, VMS, AIS, Satellite-AIS (figure 2), VTS, call-in regimes) to observation systems (coastal radar, satellite imaging as VDS – Vessel Detection System presented in figure 3), and assessing the feasibility for operational use by national authorities in the Horn of Africa region and cost-benefit aspects. While the use of reporting systems is in large part related to legal aspects, the challenge for experts in satellite imagery analysis is to overcome technological hurdles, taking into account its particularities, viz. its capability to detect but not identify ships, its ability to survey even the most remote regions but not to continuously monitor, and the delays between the observation of the scene and the delivery of analysis results.

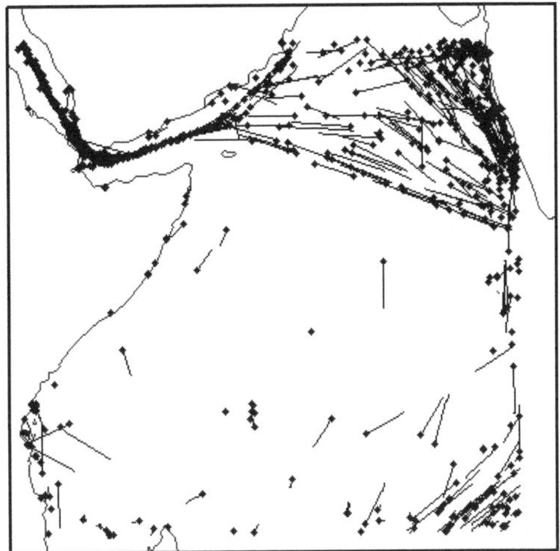

Figure 2: Sat-AIS "ExactEarth" tracks per day. "Result produced by JRC. Includes copyrighted material of exactEarth Ltd. All Rights Reserved." - 11 Dec 2012

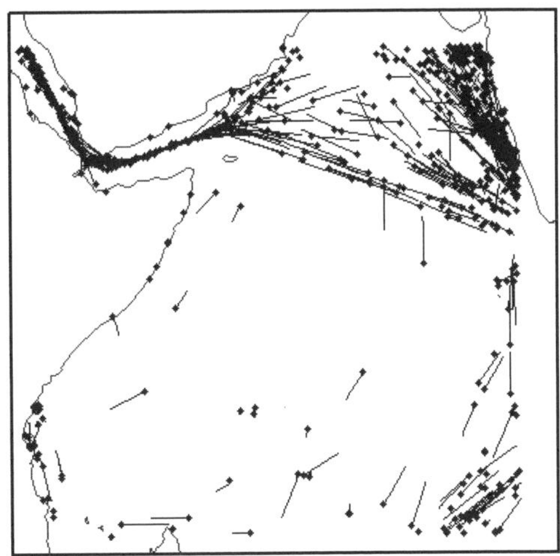

Figure 2: Sat-AIS "ExactEarth" tracks per day. "Result produced by JRC. Includes copyrighted material of exactEarth Ltd. All Rights Reserved." - 12 Dec. 2010

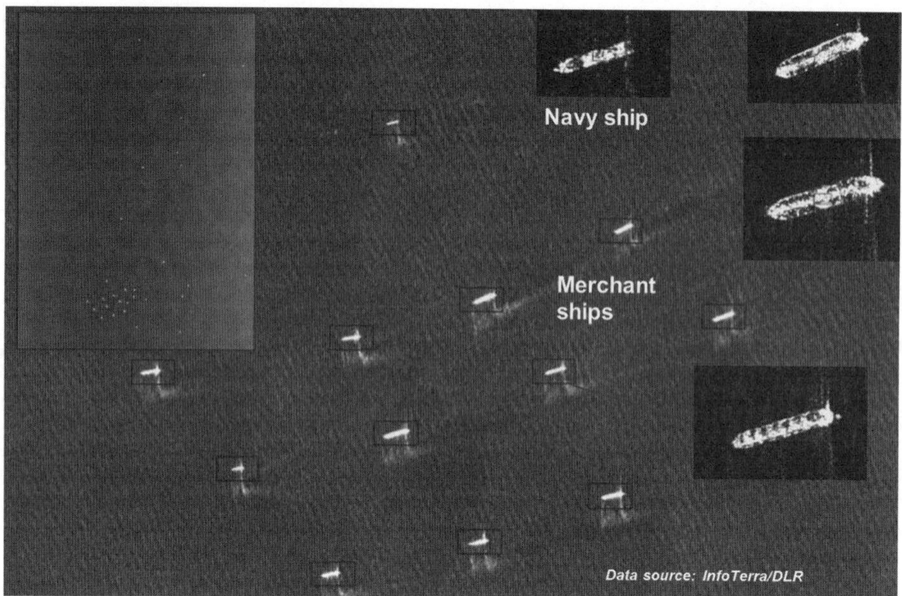

Figure 4: Gulf of Aden, 5 Dec 2010 "TerraSAR-X StripMap -Result produced by JRC. TerraSAR-X image © DLR/Infoterra 2010."

4 CONCLUSION

Writing just after the so-far-called jasmine revolutions of Tunisia and Egypt, with protests emerging in force across northern Africa and much of Arabia, it is both difficult to imagine the regional circumstances five months from now when the TransNav symposium is held in Gdynia and to view a problem such as piracy in an historic maritime zone such as the Indian Ocean, not to mention the Bab el Mandab, even the Red Sea, without giving thought to extra-scientific matters. Therefore this paper, before introducing the MARSIC and EC-JRC projects, takes time to discuss aspects of piracy that may be relevant and, certainly in the case of the MARSIC project, even decisive regarding the success of the projects. Currently, the MARSIC project's focus is on the RMI to be established in Sana'a, where volatile protests and counter-protests are occurring daily, while elsewhere in the country, aside from supportive protests in other cities, a vigorous separatist movement is underway.

Certainly both projects, from the social sciences point of view, seem to be on the right track in encouraging regional autonomy. The technology is available, through satellites, AIS, VTS, etc., to at the very least reduce the extent of the problem in the long run. And given current and past circumstances, piracy is least problematically approached by regional stakeholders—the one question being to what extent they believe that they are indeed stakeholders. Certainly cooperation between the countries of the region and Europe (and other developed nations), the sharing of technology and expertise and the financial support on the part of the EC is for the nonce as good a place to start as any.

ACKNOWLEDGEMENT

Special thanks to Dr. Harm Greidanus, EC-JRC for technical contributions.

REFERENCES

Braudel, Fernand, The Mediterranean, and the Mediterranean World in the Age of Phillip II, 2 Vol.; University of California Press, Berkeley, 1995

Braudel, Fernand, The Mediterranean in the Ancient World; Penguin Books, London, 2001

Chaudhuri, K.N., Trade and Civilisation in the Indian Ocean, and Economic History from the Rise of Islam to 1750; Cambridge University Press, Cambridge, UK, 1985.

Das Gupta, The World of the Indian Ocean Merchant, 1500-1800; Oxford University Press, Oxford, UK, 2001

ExactEarth- ExactAIS- Satellite AIS, www.exactearth.com , 2011

Ferraro, G., Baschek, B., de Montpellier, G., Njoten, O., Perkovic, M., Vespe, M. "On the SAR derived alert in the detection of oil spills according to the analysis of the EGEMP", Mar. Pollut. Bull., Vol. 60, Issue 1, Pages 91-102, doi:10.1016/j.marpolbul.2009.08.025, Jan. 2010

Floor, Willem, The Persian Gulf, a Political and Economic History of Five Port Cities, 1500-1730; Mage Publishers, Washington, D.C., 2006

Gaullier, H., Marsic 1st Monitoring Report, France Coopération Internationale, 2010

Greidanus, H., PMAR: Pilot Project on Piracy, Maritime Awareness and Risks, Trial 2010, Status 30 Jan 2011

Guardian, 'Pirate fishing causing eco disaster and killing communities, says report', June 8, 2009. Guardian, London.

Hall, Richard, Empires of the Monsoon: A History of the Indian Ocean and Its Invaders; HarperCollins, New York, 1998
Hari, Johann, 'You are being Lied to about Pirates', The Independent, London, January 5, 2009
Harsch, Rick, A Circumnavigation through Maritime History; due 2011, Maritime University of Szeczin.
Hourani, George F., Arab Seafaring; Princeton University Press, Princeton, 1995.
Info Terra & DLR, Gulf of Aden, 5 Dec 2010 "TerraSAR-X StripMap", 2010
Maritime Security -special, 2010, http://www.em-defence.com/PDFs/Maritime_Security_Special.pdf
Perkovic, M., Greidanus, H., Müellenhoff, O., Ferraro, G., Pavlakis,P., Cosoli, S., Harsch, R., "Marine Polluter Identification: Backtracking with the Aid of Satellite Imaging", Fresenius Environmental Bulletin, PSP Volume 19 – No.10b. pp. 2426-2432, Oct. 2010
Somalian piracy; Threat Map 2005-2010, http://upload.wikimedia.org/wikipedia/commons/7/7e/Somalian_Piracy_Threat_Map_2010.png

Health Problems

26. Systems for Prevention and Control of Communicable Diseases on Ship

C. Jerome
State University of New York Maritime College, New York, USA

ABSTRACT: The paper reviews developments in systems for prevention and control of communicable diseases on ships. After discussing strengths and weaknesses of existing systems, the paper proposes an approach to improve both prevention of disease occurrence and response to occurrences of illness – of both uncertain and confirmed diagnosis. The paper discusses specific measures for pre-embarkation, during voyage, prior to disembarkation and after disembarkation.

Communicable diseases pose special problems and probabilities at sea. A number of factors increase the risk of infection and contagion:

- Many strangers at close quarters
- People from various geographical and social backgrounds, with great variation in pathogens carried and in immunities
- Stagnant air in closed rooms
- Animal carriers – mosquitoes, rodents, birds
- Limited water and food supplies at risk to contamination from sea air and water as well as from shipboard organic waste
- Common ventilation system for a large population
- Spas, pools and buffet-misting devices that generate aerosols
- Hard-to-track exposures in visits to and from ports and other ships

Some of the major infectious disease concerns on ship are: yellow fever, malaria, gastro-intestinal, respiratory, sexually transmitted, chicken pox, measles and rubella.

1 THE LAST FEW DECADES

Several developments during the past few decades have made on-ship disease prevention and control systems more significant:

- An increase in seagoing crew and passengers.
- An increase in size of ships
- A greater threat of purposeful spread of communicable disease
- The international AIDS epidemic
- The emergence of Severe Acute Respiratory Syndrome (SARS)
- The H1N1 Influenza pandemic.

At the same time, international cooperation, advances in medical knowledge, technological capabilities for information management, communication and rapid delivery of supplies have improved chances for preventing and controlling communicable disease on ship. The U.S. developed VSP (Vessel Sanitation Program) in the early 1970s (Centers for Disease Control 2005). Canada initiated its CSIP (Cruise Ship Inspection Program) in 1998 (Saginur et al. 2005). In 1999 the European Union initiated work on a system for the control and prevention of influenza-like illness on ship (Commission of the European Communities 2000). In 2005 the World Health Assembly agreed upon revised World Health Regulations (World Health Organization 2008), promoting global surveillance and detection measures against public health emergencies. In 2006 the EU launched the SHIPSAN Project, and in 2008 SHIPSAN TRAINET for the purpose of controlling on-ship communicable diseases (Mouchtouri et al. 2009).

Standard procedures and systems for various activities within the field of on-ship communicable disease control and prevention have emerged. Inspection procedures and systems for preventing infectious diseases and responding to outbreaks have been established in many countries. Centers for Disease Control (2005) and Saginur et al. (2005). give excellent descriptions of these operations. Mouchtouri & Westacott et al. (2010) and Mouchtouri & Nichols et al. (2010) survey on-passenger- ship diseases and inspection practices and systems, and they discuss work on standardization and internationalization.

However, the field is only beginning to consider a more unified systems view. Canals et al. (2001), Hasanzadeh et al. (2005) and Jeremin (2009) do emphasize the importance of integration of services and components. Armitage et al. (2009) highlight the need for healthcare in general to pursue integrated models and systems:

> ...It is important for decision makers and planners to choose a set of complementary models, structures and processes to create an integrated health system that fits the needs of the population across the continuum of care.

Jensen et al. (2010) applies the same thinking specifically to health of seafarers, pointing out inefficiencies in the compartmentalized state of various practices.

With similar concerns in mind, I offer a conceptual-level outline for a distributed multi-ship communicable disease prevention and control (CDPC) system. The goal is to integrate services and component systems, facilitating efficiency and synergies.

I also make a few suggestions for handling particular difficulties.

2 A CONCEPTUAL VIEW OF A CDPC SYSTEM

I am describing not a currently functioning system, but rather, a conceptual view that incorporates the multiple facets of existing subsystems, operations and current technological capabilities. From such a vantage point, I suggest approaches to a couple of issues. Such a system's purpose is, as its name indicates, to prevent and control communicable diseases. What exactly do we mean by "the system"? Generally I'll be referring to the set of hardware and software components collaborating for this purpose. However, at certain points it will be essential to discuss the roles of the human components.

The first steps to manage the complexity of such a system are to identify the constituents (≡ stakeholders ≡ actors) and to list the use case packages (≡ categories of the system's interaction with constituents).

2.1 *Constituents*

The system's constituents are passengers, crew, ship company, ship repair/maintenance contractors, ports, other transportation units (e.g. helicopter services), external organizations (maritime, technology, environmental); friends and family; government regulators, suppliers; law enforcement; systems team.

I am using *crew* to include both on and off-ship personnel (including the health and medical team) working to operate the ship(s) safely.

2.2 *Use case packages*

2.2.1 *Access library*

Provide fast and reliable access to information (symptoms and syndromes, disease occurrence probabilities and confirmation, prevention and response) about communicable diseases at sea. It is not necessary that the CDCP system have its own library. What it does need to do is manage the library interface.

2.2.2 *Maintain non-human resources*

Inspect, test, monitor and repair resources and conditions:

- Environment
- Infrastructure and equipment
- Supplies and materials
- HVAC systems
- Waste and recycling
- Animals (including pests)

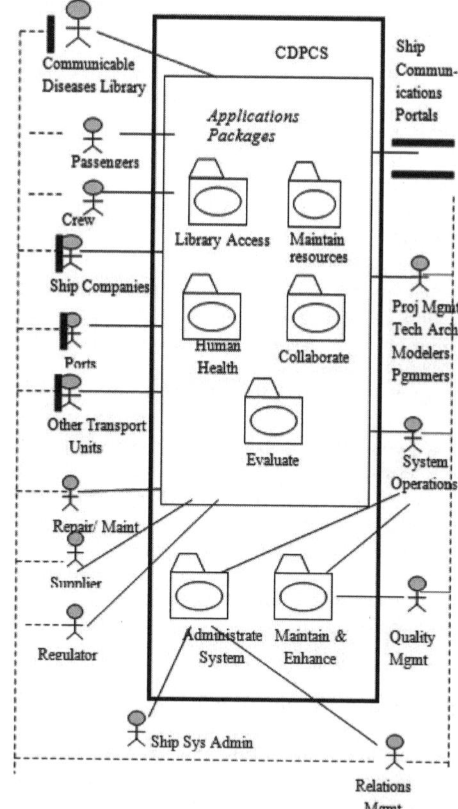

Figure 1. Context Diagram

2.2.3 *Promote human health*

This includes pre-embarkation information gathering, vaccinations, medical examinations, prevention, treatment of on-board occurrences of symptoms, syndromes, diseases and post-disembarkation tracking, treatment and outcome tracking.

2.2.4 *Collaborate*

Work with off-ship parties concerned with communicable diseases. Maritime, travel, medical organizations, port authorities, other transportation means, government (local, national and international) entities, NGOs, suppliers need to participate.

2.2.5 *Analyze, evaluate, plan*

- Maintain history.
- Evaluate past prevention, treatment and control activity.
- Analyze risks and scenarios.
- Plan for development and for contingencies.

2.2.6 *Administrate system*

Help Desk, assign User IDs, credentials. Maintain constituent information, send news alerts.

2.2.7 *Maintain and enhance system*

This includes all activities of the systems teams:

- relations management
- project leadership/management
- quality management
- operations/environment
- technical architecture
- analysis/design/modeling
- implementation.

2.3 *Context Diagram*

A context diagram (Figure 1) visually summarizes the constituent and use case relationships.

3 TECHNICAL ARCHITECTURE

A multi-ship system needs a distributed architecture. Each ship would both keep and transmit to a central server records of symptoms, syndromes, test results, diagnoses and treatments. (Deltas between on-ship and central server person information are sufficiently complex to need business-rule negotiation.) Center-based programs can rapidly analyze occurrences and patterns over a whole area. Figure 2 shows an overview of the technical architecture.

The central server connects to an infectious disease library and other information sources. With an ongoing connection, each ship can receive and convey alerts, as well as obtain probabilities, diagnoses, and suggested procedures. (Lu 2009 describes a J2EE and GIS-based method for displaying geographical patterns of infectious diseases.)

Together, the constituent list, use case partition and technical architecture give basic structure and organization for a CDCP system.

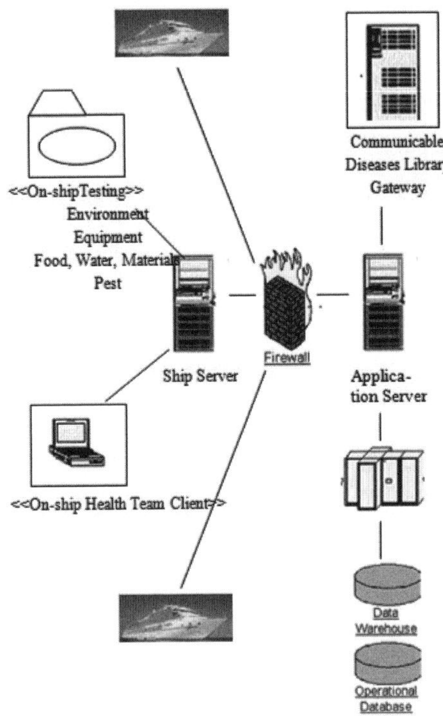

Figure 2. Technical Architecture

4 COST / BENEFIT AND FEASIBILITY

The major components of the technology indicated is implementable at present. While the size of the CDPC system costs and benefits will depend on the specifics of the implementation, we can still see the sources of incremental costs and benefits from current practices:

4.1 *Incremental costs*

- System development. This will of course not stop after the first deployment, but continue through successive versions and releases.
- System maintenance and ongoing evaluation. The system will require responsible people both on individual ships and at central locations. As usual with distributed systems, the relative cost per ship [person, person-kilometer] will decrease as the number of ships [person, person-kilometer] increases, because of the sharing of the costs.
- Educational and cultural activities. These will play a key role in assuring prevention and control of accidental or purposeful contagion (see section 5).

(I have not mentioned costs of health personnel and operations, or of international collaboration, because these costs will be needed with or without such a system.)

4.2 *Incremental benefits*

- Better organization and rapid communication of medical knowledge and health-related data will speed prevention of, and response to, infectious disease.
- Long-distance medical consultation and treatment will avoid many otherwise necessary ship diversions. [This is the point stressed by Patel (2000).]
- Standardization of procedures will enable increased quality of service.
- Technology and a globalized view of spread of disease will facilitate rapid deployment of appropriate medical methods.

As with costs, the amount of benefit – in both human and financial terms – will be dependent on the extent of collaboration among maritime entities.

5 SOME CHALLENGES

Two challenges for CDPC systems are in the areas of traveler cooperation and system and resource protection.

5.1 *Traveler cooperation*

The problem of getting people (both passengers and crew) to follow recommended practices has long been a trouble spot in controlling on-ship and post-disembarkation diseases. Merson (1975) reported that 75% of occurrences of gastroenteritis (as reflected in his questionnaires to passengers) were unreported in ship logs. Saginur et al. (2005) mention similar problems with regard to sexually transmitted disease. Neri et al. (2008) detail various ways people fail to cooperate – from not washing hands to not reporting illness (occurring prior to embarkation or during voyage) to not using hand sanitizer.

To increase traveler cooperation with disease prevention and control, we need to look at the human components of our CDCP system. A relations management team can encourage, educate and win hearts and minds of passengers and crew. (Centers for Disease Control (2005) mentions certain individuals on ship whom the passengers will feel free to consult, while Saginur et al. (2005) also suggest counseling for passengers.) We need to see this as an ongoing mobilization: discussions and cultural activities starting prior to disembarkation and continuing long after disembarkation. Encouragement needs to be given to communication and artistic events having themes like health, body and people's lives on ship. In doing this, there is a risk of too much – evoking non-interest or even resentment – as well as of too little. Activities need not be too narrowly focused. One example: a shipping line might run a short playwriting contest with the theme, Health On Ship. Accepted contest entries could be performed (either off-book or as staged readings) on ship, with audience feedback.

The relations management team needs to get to know everyone, to keep track of people's preferred communications means and languages, and to encourage people's concerns with others' health.

5.2 *System and resource protection*

Protection involves both safety – measures against accidental harm – and security – protection from intentional harm. The subject ranges over questions of backup, recovery and disaster recovery, identity and authentication, authorization, privacy and various types of defense. Here I discuss two of these areas:

- access rights to system information and processes
- bioterrorism.

5.2.1 *Access rights*

For a simple system, access rights may be set up by the ship system administrator or system operations personnel, with the rights – e.g. read, write, execute – indicated (à la UNIX or Microsoft's Right Management Services) in a central directory and/or attached with a simple indicator to each document, component or program. However, with more complicated decisions to be made about who can do which operations, a more sophisticated management of rights is necessary. We need to introduce an authorization mechanism to deal with complex access actions requested, under various ship alert conditions, by actors with differing credentials. Ritter et al (2006) give an access rights model for HVAC systems, but the various actions for which rights are required are rather limited. Our CDCP systems need to deal with more complex situations: An example: A researcher may want to check the backgrounds of travelers in order to evaluate factors in responsiveness to a certain treatment. The system needs to allow him/her to access only certain information. The investigator may have credentials sufficient for modifying some of the information in certain limited ways. Such rights may also depend upon the ship alert conditions at the time. This kind of system needs a more flexible model that sets up as separate types: credentials, resource gateway, permission rule, along the lines of the type model shown in Figure 3.

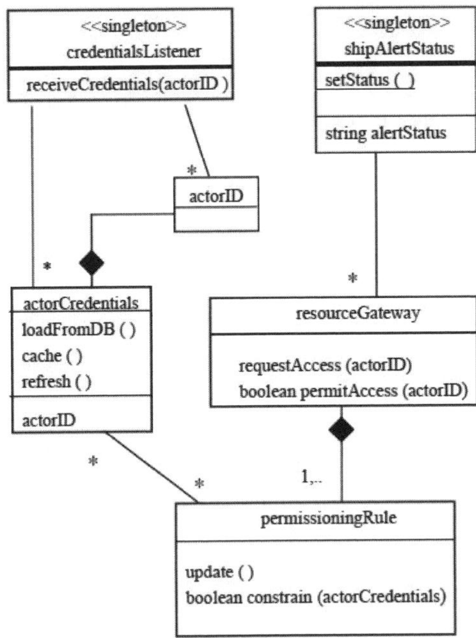

Figure 3. Access Rights Administration Type Model

Credentials need to be loaded and cached when an actor logs into the system and updated on an event-driven basis. Resource gateways, part of the interface to restricted resources/operations check permission rules and decide whether to grand a requesting actor access to a resource/operation.

5.2.2 *Bioterrorism*

Bravata & McDonald et al. (2004) review a large number of systems for processing surveillance data for bioterrorism-related constellations – some that are general CDCP systems and others specifically focusing on bioterrorism – and find them somewhat disappointing. Some of their disappointment is in regard to

- insufficient automation
- lack of documentation and evaluation
- scarcity of reference information for these systems.

These are certainly reasons for urging more development effort. However, Bravata & Sundaram et al. (2004) are also disconcerted by the high number of false positives produced by the detection systems they review. Their conclusions need further consideration.

Where incidence of dangerous attacks is low, an easy-to-administer screening test with high sensitivity and reasonable specificity is not a final adjudicator, but an excellent start at checking for such attacks. The Bravata & Sundaram et al. example with reference to the anthrax attacks of 2001 will serve us here:

> The Trenton, New Jersey, state police evaluated > 3,500 samples of suspicious powders, and none contained anthrax... For the purpose of illustration, let us assume that, before testing, 5 of these 3,500 samples were estimated to contain anthrax (i.e., pretest probability equals 0.0014). If a detection test had a sensitivity of 96% and specificity of 94% (i.e., the lower range reported for SMART/ALERT), we can calculate the post-test probability of anthrax with both positive and negative test results by using Bayes' theorem... If such a detection system indicated a positive result, the probability that the sample contained anthrax would be approximately 2%. That is, 98% of the positive results would be false-positives. If the system indicated a negative result, the probability of anthrax in the sample would be 0.006%. Thus, the test would be useful when negative, but provide little help if positive.

The value of a screening test (given that its specificity is fairly high – as is the case in the example) is not that it conclusively identifies an attack, but rather, that it significantly and safely reduces the number of cases to be investigated. Let

A_a = {sample contains anthrax}
A_n = {does not contain anthrax}
B_+ = {sample tests positive}
B_- = {tests negative}.

The probability of B_+, in the example, is given by:

$$P(B_+) = P(A_a)\,P(B_+|A_a) + P(A_n)\,P(B_+|A_n) \qquad (1)$$

From (1) we have:

$$P(B_+) = (5/3500) \times 0.96 + (3495/3500) \times 0.06$$

$$= 0.0612.$$

Thus the detection test would show positive for .0612 x 3500 = 215 samples. As long as the test is inexpensive and easily administered, this provides a tremendous saving of effort. Further investigation could proceed for the 215 positives, rather than for the entire 3500. To demand more from one automated test is perhaps unrealistic, perhaps risky. Whatever the test's reading, it generally would be advisable, before proceeding with drastic measures, to investigate the positives further.

Where do we go after an initial screening test? Yes, we can and must do more testing. Eventually, however, experienced judgment is needed to decide upon emergency measures. Also, we must:

- encourage travelers to be alert to danger signs
- discourage mass paranoia
- pool people's knowledge.

This takes old-fashioned human expertise, spirit-building talk whether in person or over electronic social networks or by other means. Thus, just as in our discussion of traveler cooperation, we arrive at developing the human components in our CDPC system.

6 CONCLUSION

A systems approach to the heretofore somewhat disparate activities in prevention and control of on-ship communicable diseases will:

- strengthen effectiveness and efficiency,
- encourage standards,
- promote synergies.

This approach entails developing both the technical and the human components of a Communicable Diseases Prevention and Control system. I have presented a conceptual model for such a system.

We can encourage passenger cooperation by promoting discussion and cultural activity.

A CDPC system needs a sophisticated access control model, incorporating a number of types: resource gateway, ship alert status and permissioning rules.

Guarding against bioterrorism can be made more efficient with easily administered first-stage screening tests, followed by investigation of positives. Beyond testing, we must cultivate a conscientious spirit among travelers.

REFERENCES

Armitage G.D., Suter E. & Oelke N.D. 2009. Health systems integration: state of the evidence. International Journal of Integrated Care, 9(17 June 2009).

Bravata D.M., McDonald K.M., Smith W.M., Rydzak C., Szeto H., Buckeridge D., Haberland C. & Owens D.K. 2004. Systematic review: surveillance systems for early detection of bioterrorism-related diseases. Annals of Internal Medicine. 140:910-922.

Bravata D.M., Sundaram V., McDonald K.M., Smith W.M., Szeto H., Schleinitz M.D. & Owens D.K. 2004. Detection and diagnostic decision support systems for bioterrorism response. Emerging Infectious Diseases. 10(1):100-108.

Canals M.L., Gómez F. & Herrador J. 2001. Maritime health in Spain: integrated services are the key. International Maritime Health. 52(1-4): 104-116.

Centers for Disease Control. 2005. Vessel sanitation progam operations manual. Atlanta: National Center for Environmental Health.

Commission of the European Communities. 2000. Commission decision of 22 December 1999 on the early warning and response system for the prevention and control of communicable diseases under decision No 2119/98/EC of the European Parliament and of the Council. (2000/57/EC). Official Journal of the European Communities L(21):32-35.

Hasanzadeh M.A., Azizabadi E. & Alipour N.A. 2005. Health services system for seafarers and fishermen in Iran. International Maritime Health. 56(1-4): 173-184.

Jaremin B. 2009. Strategies, means and models of health care in the Polish maritime industry, 1945-2007: summary of the presentation made at the time of the First International Congress of Maritime, Tropical, and Hyperbaric Medicine held on 4-6 June 2009 in Gdynia, Poland. International Maritime Health. 60(1-2): 75-76.

Jensen O.C., Lucero-Prisno, D.E. & Canals M.L. 2010. Integrated occupational health care for seafarers across the continuum of primary, secondary and tertiary prevention. International Journal of Integrated Care. 10(8 March).

Lu X. 2009. Web GIS based information visualization for infectious disease prevention. Intelligent information technology application, Third international symposium. 3: 148-151.

Merson M.H., Hughes J.M., Wood B.T., Yashuk J.C. & Wells J.G. 1975. Gastrointestinal illness on passenger cruise ships. Journal of the American Medical Association. 231(7): 723–727.

Mouchtouri V.A., Black N., Nichols G., Paux T., Riemer T., Rjabinina J., Schlaich C., Menel Lemos C., Kremastinou J., Hadjichristodoulou C. & SHIPSAN TRAINET project. 2009. Preparedness for the prevention and control of influenza outbreaks on passenger ships in the EU: the SHIPSAN TRAINET project communication. Euro Surveillance. 14(21): pii=19219.

Mouchtouri V.A., Nichols G., Rachiotis G., Kremastinou J., Arvanitoyannis I.S., Riemer T., Jaremin B., Hadjichristodoulou C., for the SHIPSAN partnership. 2010. State of the art: public health and passenger ships. International Maritime Health. 61(2): 49-99.

Mouchtouri V.A., Westacott S., Nichols G., Riemer T., Skipp M., Bartlett C.L.R., Kremastinou J., Hadjichristodoulou C. & SHIPSAN partnership. 2010. Hygiene inspections on passenger ships in Europe - an overview. BioMedCentral Public Health 10(122).

Neri A.J., Cramer E.H., Vaughan G.H., Vinjé J. & Mainzer H.M. 2008. Passenger behaviors during norovirus outbreaks on cruise ships. Journal of Travel Medicine. 15: 172–176.

Patel, Toral. 2000. A cost-benefit analysis of the effect of shipboard telemedicine in a selected oceanic region. *Journal of Telemedicine and Telecare*. 6: 165-167.

Rachiotis G., Mouchtouri V.A., Schlaich C., Riemer T., Varela Martinez C., Nichols G., Bartlett C.L.R., Kremastinou J., Hadjichristodoulou C. & SHIPSAN partnership. 2010. Occupational health legislation and practices related to seafarers on passenger ships focused on communicable diseases: results from a European cross-sectional study (EU SHIPSAN PROJECT). Journal of Occupational Medicine and Toxicology. 5(1): 1.

Ritter D., Isler B., Mundt H.J. & Treado S. Access control in BACnet. 2006. American Society of Heating, Refrigereration and Air-Conditioning Engineers Journal. 48 (November): B26-32.

Saginur R., Birk H. & Committee to Advise on Tropical Medicine and Travel (CATMAT). 2005. Statement on cruise ship travel. Canada Communicable Disease Report. 31 / (ACS-8).

World Health Organization. 2008. International Health Regulations (2005) 2nd edition. Geneva. World Health Organization.

Maritime Ecology

27. Coastal Area Prone to Extreme Flood and Erosion Events Induced by Climate Changes: Study Case of Juqueriquere River Bar Navigation, Caraguatatuba (Sao Paulo State), Brazil

E. Arasaki
Escola Politécnica da USP, Sao Paulo, Brazil;Instituto Nacional de Pesquisas Espaciais, S.J.Campos, Brazil

P. Alfredini
Escola Politécnica da USP, Sao Paulo, Brazil; Instituto Mauá de Tecnologia, Sao Caetano do Sul, Brazil

A. Pezzoli & M. Rosso
Politecnico di Torino, Turin, Italy

ABSTRACT: According to the IPCC, the forecast for the year 2100 is an increasing of global average temperature, whose impacts in winds, waves, tides, currents and bathymetry will produce real risks of extreme events due to climate changes. Juqueriquere River is Sao Paulo State (Brazil) North Coastline major waterway. Due to minimum channel depths in the coastal bar, navigation is only possible for small leisure crafts and fishing boats and some cargo barges during higher tidal levels. This study case has been evaluated according to the relative sea level and wave climate scenarios forecasting, based on the meteorological recognition patterns of the last decades data for tides and waves. The impact of climate changes is obtained from this knowledge. The main goal of this paper is to have the initial conceptual description about the impacts on the bar navigation conditions of Juqueriquere to obtain guidelines for master nautical plans.

1 INTRODUCTION

The Juqueriquerê Catchment is the major in São Paulo State (Brazil) North Coastline (Fig. 1). The Juqueriquerê Waterway is a 4 km estuarine channel used by small piers and docks. The entrance bar doesn't have any amelioration works and the boats maneuvers are difficult and dangerous. Beyond environmental impacts, the cost - effective improvements consists to bar jetties calibration, this solution means to talk about costs of 5 M €, or the permanent maintenance with local dredging works, which costs, in the long term of decades, will be the same.

According to the IPCC forecasting, there is the awareness that conditions of bathymetry, tides, winds, currents and waves for next decades shall have climate changes impacts. The project goal is to overcome the contraposition that it emerges between the defence against the hydraulic risk and the management to preserve the environmental protection for nautical purposes. The risk is understood, in a qualitative way, as composed by Hazard, Exposure and Vulnerability (Kron 2008).

2 SOME HISTORY

Example of navigation possibilities in the waterway was the dock operation of the English Lancashire General Investments Company, owner of the Blue Star Line Navigation Company. In the period 1927 - 1967, the “Fazenda dos Ingleses” (English Farm), has sent the tropical fruits production to England. The railway line of the farm had 120 km, with docks and warehouses in the right bank of the estuary. The Packing House, in the dock area, was considered the second of this type in South America. The cargo boats, more than 20 in the forties decade, had an individual load of 55 dwt (Fig. 2). In March 1967, a strong debris-flow, rain more than 600 mm in 2 days (monsoonal rates), combined with storm surge, caused more than 400 casualties and material losses in Caraguatatuba. After that, the docks were closed (Arasaki 2010).

In the last four decades, the nautical purposes of leisure and fishing boats increased. It is important to mention the recent interest as supplier area for the offshore LNG and oil. The plant for the gas treatment is located in the left bank of the river and many of the facility heavy cargo equipments used large barges push-pulled by tugs (Fig. 3).

3 TIDE VARIABILITY

Considering the CDS (Companhia Docas de Santos) datum, extreme LLW level, in Figure 4 are presented our study conclusions about São Paulo State coastline tidal variability for the last six decades (1944 - 2007). A consistent linear response shows: 1. Overall period: rising rates (cm/century) for MSL (23.2), HHW (36.5) and LLW (41.8); 2. Period before 1969: 1.1, - 7.3 and 54.3 and 3. Period after 1975: 40.9, 44.9 and 75.4.

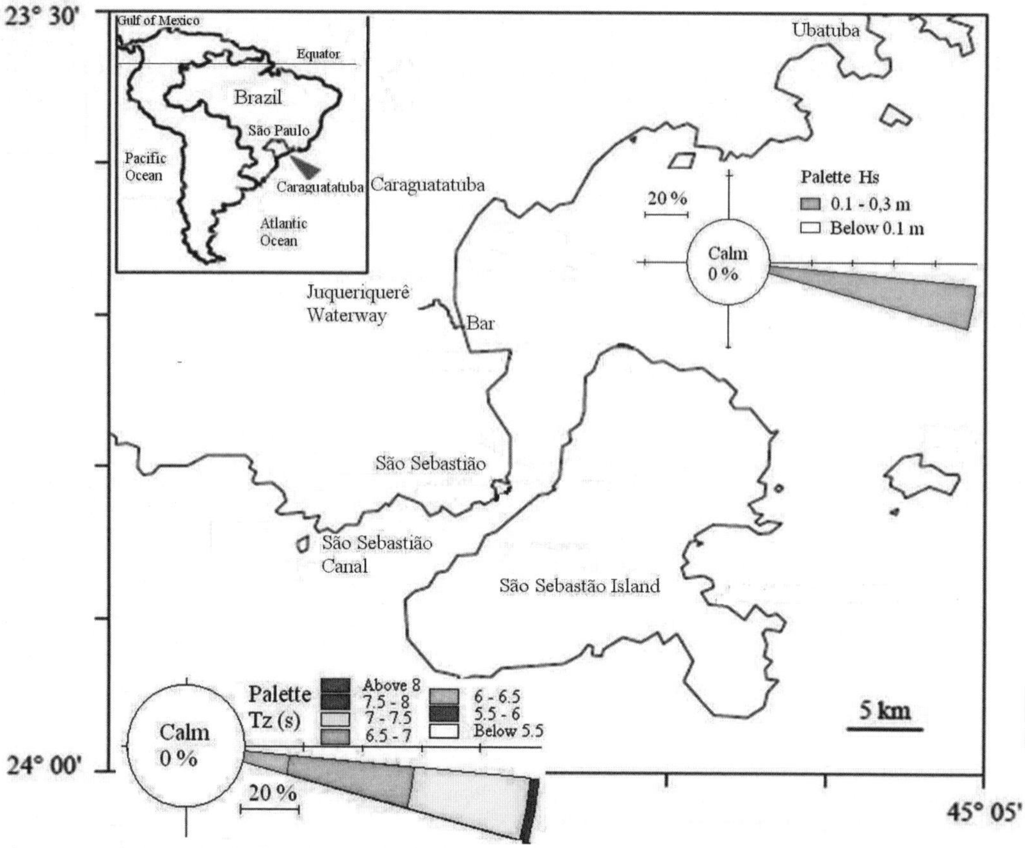

23° 30'
Gulf of Mexico
Equator
Brazil
São Paulo
Pacific Ocean
Caraguatatuba
Atlantic Ocean
Caraguatatuba
Ubatuba
Palette Hs
0.1 - 0,3 m
Below 0.1 m
20 %
Calm 0 %
Juqueriquerê Waterway
Bar
São Sebastião
São Sebastião Canal
São Sebastão Island
Palette Tz (s)
Above 8
7.5 - 8
7 - 7.5
6.5 - 7
6 - 6.5
5.5 - 6
Below 5.5
Calm 0 %
20 %
5 km
24° 00'
45° 05'

Figure 1. Location map with significant height and average period local wave roses.

Figure 2. Boats moored at the quay, dockyards warehouses and railway terminal (1940 decade) of the English Farm.

Figure 3.Heavy cargo equipments using large barges push-pulled by tugs (February 2010).

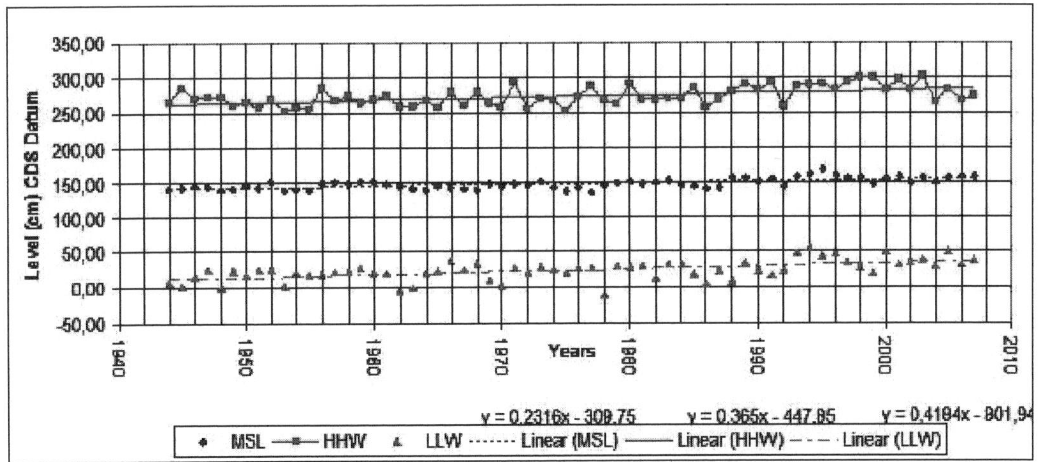

Figure 4. Annual tidal levels variability (1944 – 2007) for São Paulo State (Brazil) coastline.

4 FLUVIAL MORPHOLOGY

The Juqueriquerê Catchment has the following main features: area of 430 km^2, long term average discharge of 11 m^3/s, heavy rainfall rates (around 3000 mm/year) producing high fluvial sediment transport, floods and debris-flows. The last ones are due to the steep slopes and the altitude (~ 1000 m) of the Serra do Mar mountains near the coast, producing the orographic effect, which rapidly condensates the sea humidity.

The fluvial dynamics is of high solid transport capacity and fluvial and coastal morphology transformations, combined with recurrent and intense flood events that cause extensive risks and damages to population and infrastructures, causing riparian and coastal region with important anthropic impact.

Strong debris-flows occur in this region, because events similar to monsoonal rain rates (higher than 300-400 mm per day) occur in multi decadal periods. The region history records shows this type of strong events in 1859, 1919, 1944 and the last and more catastrophic in March 1967, Figure 5 shows the damage caused in the English Farm dock yard, warehouses and Packing House (Arasaki, 2010).

5 WAVE CLIMATE

According to Marquez & Alfredini (2010), the offshore climate may be described by the ECMWF – European Centre for Medium-Range Weather Forecast – ERAS40 project. In Figure 6 are presented the H_s, significant wave height, and T_z, average period, for the nearest grid point 1958 - 2001 data series. The local wave climate may be described by the H_s and T_z roses in Figure 1, obtained from the coastal buoy data records survey of São Paulo State coastline (1982 - 1984) treated with the DHI software MIKE 21 NSW (Nearshore Waves).

6 COASTAL SEDIMENTS AND MORPHOLOGY

In general, sediment's samples near the bar show dominance of material like fine sand, silt and clay. According to Venturini (2007), the grain size proportion of clay and silt in the first 2.0 m from the bottom surface were 15 to 20% and 35 to 55%, respectively.

The waterway is strongly restricted by the maritime bar depth, 0.3 - 0.5 m according to Chart Datum (MLWS) with MSL (Z_0) of 0.7 m, and the entrance is instable, with seasonal migration. Indeed, according to Bruun (1978) criteria for overall stability, the preliminary evaluation of Ω/M_{tot}, ratio of the spring tidal prism per total littoral drift, is 5 to 10. The Ω is of the order of 1.0 Mm^3, including river discharge, and M_{tot}, based on local breaking wave climate, is around of 0.1 Mm^3/year. For ratios less than 20, there are comprehensive bars with very shallow depth, being typical bar-bypassers, and in this case depending upon the strong flushing during the rainy season.

The morphological behavior of the bar is being studied based on seasonal thalweg bathymetrical surveys (Fig. 7) correlated with meteorological conditions corresponding to the three months before accumulated rain: April 2004: 382 mm; March 2009: 688 mm; September 2009: 313 mm; February 2010: 634 mm and September 2010: 463 mm. Wave climate is also considered.

7 CONCLUSIONS

Considering the awareness about the importance of climate changes impacts in a coastal area prone to extreme flood and erosion events, important issues to support confidence, or not, to the decision of construct two jetties (rigid structures) solution, maintenance dredging (flexible solution), or non intervention in the waterway are:

- 1. There is an overall sea level rising trend, which matches with the IPCC forecasting; 2. LLW has the highest rate of linear tidal rising (75 cm/century); 3. There is an overall tidal range reduction; 4. The tidal prism will change, and the tidal currents velocity should increase, if the HHW levels will drown large fluvial areas, compensating the velocity reduction due to the tidal range decreasing; 5. Considering the issues above, the river bar depth should increase and 6. The overall rise of the sea will produce more coastal erosion and littoral drift, in opposition to the outcome of issue 5.
- It is possible to observe a general significant height and average period wave increasing for annual averaged figures over than 1.5 m and 8.0 s, and the corresponding decadal maximum wave, from 3.9 m and 13.0 s in the sixties to 4.5 m and 14.5 s in the nineties. It means increasing swell. Hence, should be a trend to increase littoral drift, reducing bar depth.
- There are some areas of mud, which may be fluid sufficient to consider the nautical bottom concept (PIANC et al. 1997), in practice for mud density lower than 1250 kg/m^3. In these cases it is possible to reduce the under keel clearance. The analysis of September 2010 and March 2011 survey, with detailed samples of the bar and bathymetry, should provide confidence for this answer.
- About the thalweg shifting migration, it is possible to conclude: 1. Like for the monsoon weather, the main channel alignment depends upon flood periods, according to rain rate of; 2. The shifting between two adjacent thalwegs may be produced by extreme river flow conditions, or a storm surge.

Awareness with climate changes impacts importance for the intervention's plan must be considered to obtain a final balanced solution among structures, dredging and non structural measures for nautical master plan.

It is important to recognize that great natural events are not avoidable, but great disasters are, as the ancient Greek Aristotle (384-322 B.C.) said, "It is probable that the improbable will happen" (Kron 2008). Unfortunately, closing this paper, we have to recognize this historical truth: in January 2011, a very large debris-flow phenomena (more than 600 mm rain rate in two days) in the Serra do Mar mountains of Rio de Janeiro State (Brazil), which border São Paulo State, killed more than 700 people, being considered by the ONU the eighth of this type since 1900.

We want to provide an intervention's plan for Juqueriquerê Waterway for the next 50 years about geomorphologic, structural and no-structural hydraulic shape in reference to: 1. Plani-altimetric historical evolution analysis of the coastal thalweg, validating migration model considering wave climate, tides and fluvial discharges; 2. Morphological analysis of bankfull, floodplains and perifluvial areas; 3. Grain size analysis of transported sediments; 4. Meteorological and oceanographic analysis (Pezzoli et al. 2004a and 2004b; Cristofori et al. 2004) 5. Hydraulic analysis by using software HEC_RAS of US Army Corps of Engineers. After, we want to draft a maintenance plan for Juqueriquerê Waterway reckoning on the basis: 1. Ordinary maintenance isn't replaceable with structures; 2. Maintenance actions must be specific, pointed and planned; 3. Maintenance works must be well-according with landscape and with the ecosystem.

Figure 5. Destruction in the dock area and Packing House of English Farm after the debris flow of March 1967.

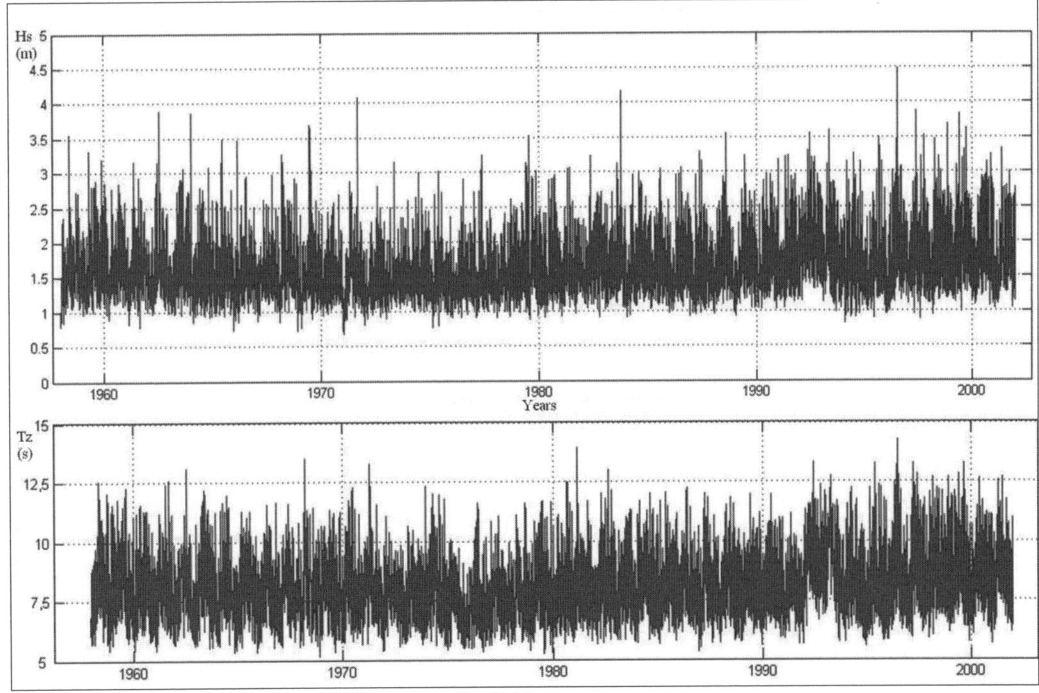

Figure 6. Deep water wave data series of H_s and Tz (1958 – 2001).

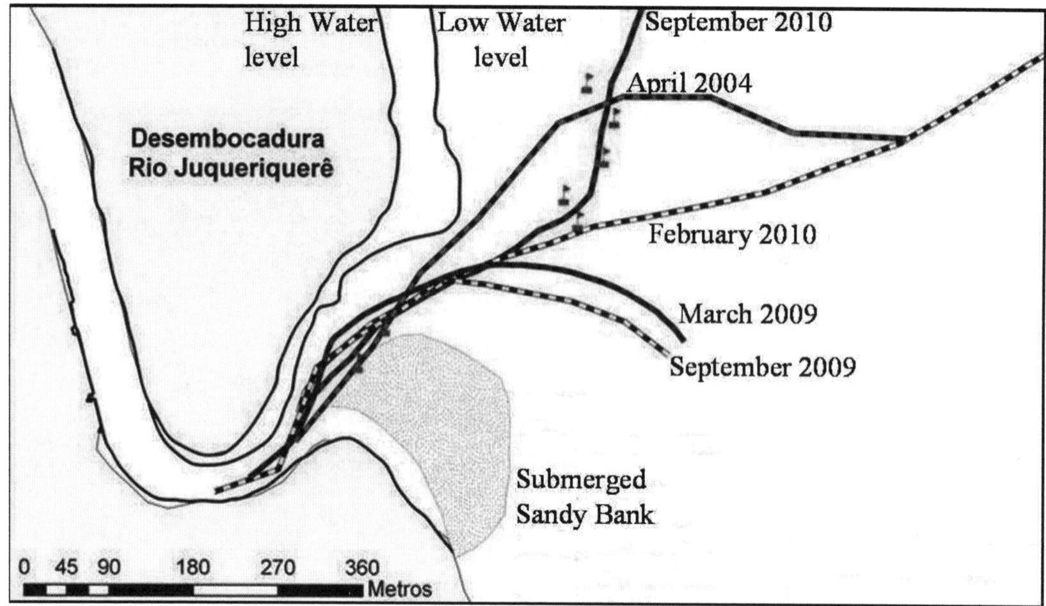

Figure 7. Entrance channel thalweg in rainy (April 2004, March 2009, February 2010) and dry (September 2009 and 2010) seasons.

8 ACKNOWLEDGEMENTS

This paper has the financial support of CAPES, Human Resources Improvement Agency of Brazilian Government. The authors also want to thank the support of São Paulo University, Politecnico di Torino, Instituto Nacional de Pesquisas Espaciais, Instituto Técnico da Aeronáutica, Instituto Mauá de Tecnologia.

REFERENCES

Arasaki, E. 2010. Capacidade de adaptação às mudanças climáticas globais: uma abordagem institucional para a gestão de recursos hídricos. São José dos Campos: Instituto Tecnológico de Aeronáutica Pos Doc Thesis.

Bruun, P. 1978. Stability of tidal inlets: theory and engineering. Amsterdam: Elsevier Scientific Publishing Company.

Cristofori, E., Pezzoli, A., Rosso, M. 2004. Analisi dell'interazione vento-corrente di marea in zona costiera: applicazione al Golfo di Hauraki-Auckland (Nuova Zelanda). 29° Congresso Nazionale di Idraulica e Costruzioni Idrauliche: vol. 3, 745-751.

Kron, W. 2008. Coasts – The riskiest places on Earth. 31st International Conference on Coastal Engineering, Hamburg. American Society of Civil Engineers.

Marquez, A.L. & Alfredini, P. 2010. Reconstructing significant wave height and peak period time series for a coast location: Vitoria/ ES Bay study case. 1st Conference of Computational Interdisciplinary Sciences, São José dos Campos, 23-27 August 2010. Pan-American Association of Computational Interdisciplinary Sciences.

Pezzoli, A., Resch, F. & Tedeschi, G. 2004a. Effetti della stratificazione atmosferica sull'interazione vento-onde in zona costiera. Memoria Academia. Scienze Torino 28: 43-67.

Pezzoli, A., Tedeschi, G. & Resch, F. 2004b. Numerical simulations of strong wind situations near the Mediteranean French Coast: comparison with FETCH data. Journal of Applied Meteorology 43-7: 997-1015.

Pezzoli, A. & Cristofori, E. 2005. Now-casting of strong Etesian wind in the Saroniko Gulf. World Water Research Program Symposium on "Nowcasting and Very Short Range Forecasting": 173-181.

PIANC - Permanent International Association of Navigation Congresses, IMPA – International Maritime Pilots Association, IALA - International Association of Lighthouse Authorities 1997. Approach channels - A guide for design. Tokyo: Joint PIANC Working Group II-30 in cooperation with IMPA and IALA.

Venturini, N. 2007. Influência da quantidade e qualidade da matéria organiza sedimentar na estrutura e distribuição vertical e horizontal das comunidade bentônicas na plataforma de São Sebastião, São Paulo, Brasil. São Paulo: Instituto Oceanográfico of USP MScThesis.

28. Study of EEOI Baseline on China International Shipping

Wu Wanqing, Zheng Qinggong, Wu Wenfeng & Yang Jianli
Dalian Maritime University, Dalian, Liaoning, China

ABSTRACT: After introducing the background and the development of GHG emission control regulations, according to the Guidelines for Voluntary Use of Ship Energy Efficiency Operational Indicator (EEOI) promulgated by International Maritime Organization (IMO), the GHG emissions of China's main international shipping was tested, as well as the EEOI and the related baseline were calculated. Based on the field data and the baselines, the factors affecting the EEOI were studied, and the related suggestions were put forward. Finally, the ship GHG emission control way was discussed.

1 INTRODUCTION

Global climate changing is one of the most serious problems faced by the human being. The sight-seeing of Himalayan peaks is changing due to the deglaciation, and the rising of sea level is threading the living of the residents on the Pacific Ocean islands. All of these alarm our society. Human being's activity, which gives out great mass of Greenhouse Gas (GHG), is the main cause of the climate changing as indicated in many researches. According to the climate model, the global average temperature will increase 1.4 - 5.8°C by the year 2100 as forecasted by the Intergovernmental Panel on Climate Change (IPCC) [1].

To flatten the global warming trend, the United Nation adopted the United Nations Framework Convention on Climate Change (UNFCCC) in 1992, which entered into force in the same year. In 2003, IMO Assembly adopted the resolution A.963 (23), which urged the Marine Environment Protection Committee (MEPC) to establish a shipping GHG emission baseline, and to develop a methodology to describe the GHG emission level in terms of GHG emission index. In 2005, MEPC 53th confirmed that CO_2 is the main greenhouse gas emitted by ships and approved the Interim Guidelines for Voluntary Ship CO_2 Emission Indexing for Use in Trials (CO_2 Index Guideline) by MEPC/Circ.471. In 2009, MEPC 59th approved the Guidelines for Voluntary Use of the Ship Energy Efficiency Operational Indicator (EEOI Guideline) by MEPC.1/Circ.684, replacing the use of former guideline [2-3].

Based on the above mentioned background, the EEOI of China international shipping was tested, and the EEOI baselines for different types of ship were set out. Based on the referred research results, the factors influencing the EEOI were studied, and the general methods to decrease the EEOI were introduced in the paper.

2 THE EEOI BASELINE OF CHINA INTERNATIONAL SHIPPING

2.1 *Data collection*

The related field data of the main China ocean shipping company were collected according to the EEOI Guideline (the reporting sheet shown as table 1), the following ships were involved:

32 bulk carriers, among which 6 are above 15000 DWT, 8 are about 70000 DWT, 10 are between 20000 DWT - 50000 DWT.

35 crude oil tankers, among which 5 are about 40000 DWT - 50000 DWT, 17 are about 60000 DWT - 70000 DWT, 3 are above 150000 DWT, and 10 are about 300000 DWT.

42 container ships, the tonnage of which varies from 10000 DWT (900TEU) - 110000 DWT (10020TEU).

23 product tankers, among which 4 are about 13000 DWT, 11 are about 42000 DWT, 5 are about 110000 DWT.

20 general cargo vessels, the tonnage of which varies from 6000 DWT - 30000 DWT, the majority of them fall between 10000 DWT - 20000 DWT.

3 chemical tankers, all are 2999 DWT.

Table 1. EEOI reporting sheet

Name and type of ship:					
Voyage	Fuel consumption			*Cargo data*	
(or day)	Fuel 1	Fuel 2	Fuel i*	Quantity	Distance
	t	t	t	t or TEU	nm
1					
2					
i*					

* i*=3,4,5...

2.2 *Data processing and results*

2.2.1 *Calculation of average EEOI for a certain ship type*

1. The average EEOI is calculated according to the equation 1, as given by EEOI Guideline.

$$EEOI_{average} = \frac{\sum_i \sum_j \left(FC_{ij} \times C_{Fj}\right)}{\sum_i \left(m_{cargo,i} \times D_i\right)} \tag{1}$$

Where j = the fuel type; i = the voyage number; FC_{ij} = the mass of consumed fuel j at voyage i; C_{Fj} = the fuel mass to CO_2 mass conversion factor for fuel j; m_{cargo} = cargo carried (tones) or work done (number of TEU or passengers) or gross tons for passenger ships; and D = the distance in nautical miles corresponding to the cargo carries or work done.

The unit of EEOI depends on the measurement of cargo carried or work done, e.g. tones CO_2/ (tons× nautical miles), tones CO_2/ (TEU × nautical miles), tones CO_2/ (person × nautical miles) etc.

2. Conversion factors of fuel to CO_2 adopt the values in table 2, as given by EEOI Guideline.

Table 2. Fuel mass to CO_2 mass conversion factors (CF)

	Type of fuel	C_F
1	Diesel/Gas oil I	3.206000
2	Light Fuel Oil (LFO)	3.151040
3	Heavy Fuel Oil (HFO)	3.114400
4	Liquefied Petroleum Gas-Propane	3.030000
5	Liquefied Natural Gas (LNG)	2.750000

2.2.2 *EEOI baseline fitting*

Fit the EEOI data got from 2.2.1 to power function, the baselines of different type ship are shown in Figures 1-6 (the data expressed by triangle were eliminated from fitting). The baseline is calculated according to the equation 2.

$$\text{Baseline value} = a \times \text{capacity}^{-c} \tag{2}$$

Where a = constant number, c = constant number.

2.2.3 *Results analysis*

The EEOI of Chemical tanker can't be power fitted because only 3 ships have been tested. The following rules can be concluded from the above data and function fitting results.

1. No matter the ship type, EEOI decreases with increase of the tonnage.

2. As to different ship type, EEOI of crude oil tanker is similar to that of bulk carrier, the baseline is fairly low. Baselines of container carrier and general cargo vessel are obviously higher than other types of ship because of the high speed of container carrier and the small tonnage of general cargo vessel.

3. The data distribution concentricity and the function matching degree differ from different ship types. Crude oil tanker and bulk carrier show better concentricity and better matching degree because of the relatively changeless voyage course. Product tanker and general cargo vessel show low concentricity and low matching degree because of the complicated course.

3 CONSIDERATION ABOUT THE EEOI

3.1 *The factors influencing the EEOI*

During data testing and processing, the following factors were found having big impact on EEOI:

Ship Loading Ratio. It can be obviously seen from the EEOI calculation equation that EEOI has a direct ratio relation with cargo mass, and an inverse ratio relation with oil consumption. The more cargo carried, the lower EEOI, the lower hazard to environment. As for the shipping company, full loading is one of the main objectives and means gaining more profits. But due to the fluctuation of the world economy and the shipping market, together with the cargo distribution characteristics of some certain ships, the ship loading ratio will differ from voyage to voyage. The average period indicator index will be influenced by the world economy.

Ship Age. Older ship age sometimes means older equipments and engines and high roughness of hull, which will normally lead to low energy efficiency and high resistance. So the EEOI of old ships will be relatively higher.

Voyage Distance. During the data processing, it was found that shorter voyage distance gives higher EEOI. The main reason lies in the high fuel consumption during arrival and departure a port.

Sea Condition. The sailing resistance is caused by water and air. Thrash at a rough sea and a windy day will get much higher resistance and show much higher EEOI.

Sailing Speed. The fuel consumption shows direct ratio to the cube of sailing speed. A high speed gives a high EEOI.

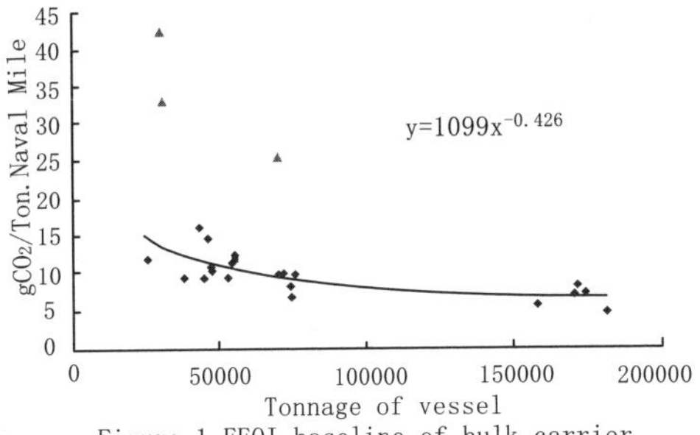

Figure 1 EEOI baseline of bulk carrier

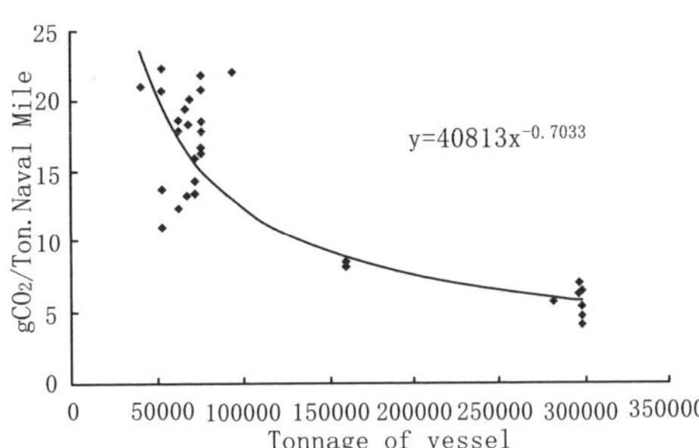

Figure 2 EEOI baseline of crude oil tanker

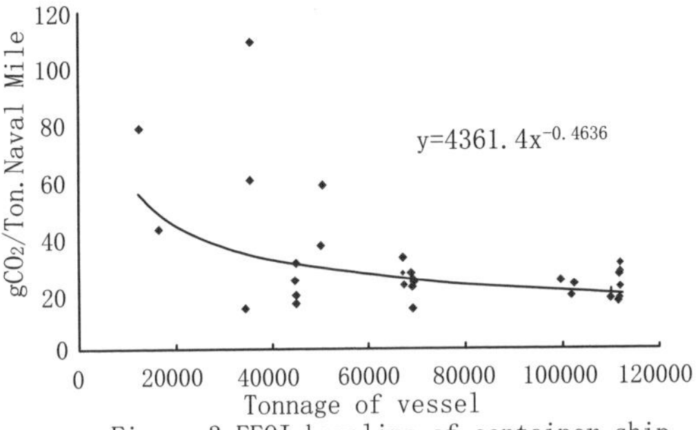

Figure 3 EEOI baseline of container ship

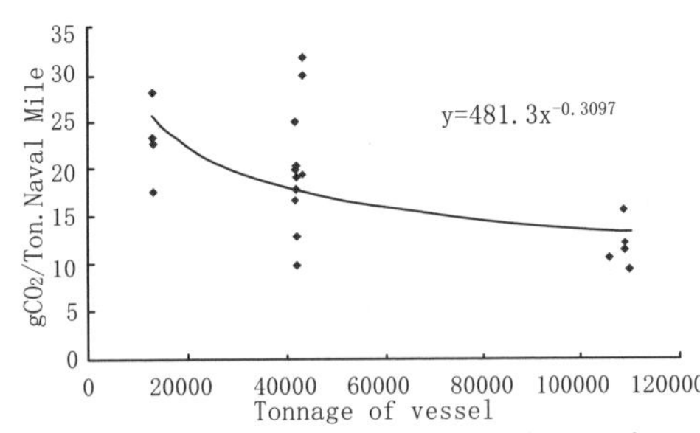

Figure 4 EEOI baseline of product tanker

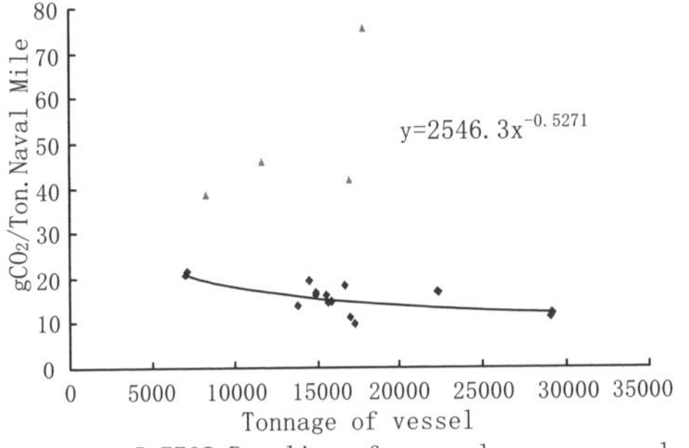

Figure 5 EEOI Baseline of general cargo vessel

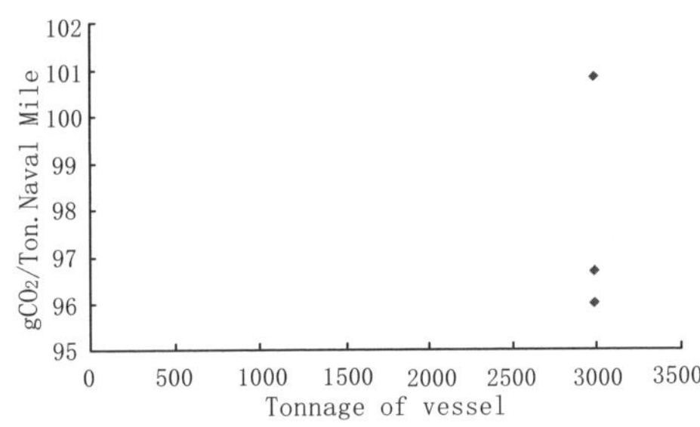

Figure 6 EEOI of Chemical tanker

3.2 *Suggestion to EEOI*

Based on the above research, the following suggestions are offered to EEOI:

1 To introduce conversion coefficient of the fuels. First, though the modern supercharging technology improved the combustion efficiency to much extent, the carbon atoms of oil will not completely convert into CO_2. It will be more serious during maneuvering or part load. Second, the fuel will inevitably be lessened during the clarify processing because of drainage and separator ejection. The lost part should be excluded during calculation.

2 To take the ship type and different cycle period into consideration and give appropriate flexibility when setting out EEOI baseline. The operation condition of different ship type is obviously different. Even of the same ship, it is different in different season or different economic environment and the period EEOI will vary to large extent. So the EEOI baseline should be given an appropriate flexibility according to the local voyage situation.

3 To consider the energy consumption and the emissions during ship building. According to the EEOI calculation equation given by EEOI Guideline, lower speed means lower EEOI. But the low EEOI at a very low speed does not mean a comprehensive high efficiency. So it would be better to introduce a comprehensive energy operation indicator which considers a relatively long period of time.

4 CONSIDERATION ABOUT SHIP GHG EMISSION CONTROL

4.1 *The choosing of GHG emission control method*

Presently the control mode of the ship GHG emission can be roughly divided into total emission control and single vessel emission control. The total control is based on Kyoto protocol. It tends to control the emission of a country or region by setting quotas. The single vessel control tends to establish an indicator similar to NOx emission taking into account the tonnage, vessel type and the rated power, but pays no emphasis to the nationality of the vessels. The former intends to achieve the final objective of GHG emission abatement. The latter emphasizes the technology means to reduce emission.

4.2 *Methods to reduce GHG emission*

Researches about GHG emission abatement have been carried out for a long time. By now the widely accepted methods mainly include technical means, operational means and market based means.

Technical means mainly include hull and propeller optimizing design, which based on the hydrodynamics [4]. Operational means include various measures other than technology, such as choosing a suitable speed and improving the fleet management.

The objective of a ship owner or a shipping company is to gain profit. Attaching additional charge to the high EEOI vessels would motivate the owners and companies to introduce new technology to reduce the emission. Based on this principle, market based means is possible.

4.3 *The fundamental way to control emission*

Although technical means, operational means and market based means have been accepted as the three main measures to reduce the emission, but as to the concrete realization there are only two methods: energy saving and energy substitution (including new energy exploitation).

Energy saving depends on the efficiency improvement, which includes improvement of combustion and thermodynamic efficiency, hydrodynamic efficiency and management efficiency.

Energy substitution mainly includes the power substitution at sea and the auxiliary electrical power substitution at berth. [5]

The energy substitution during sailing includes the using of nature gas, fuel cell, nuclear, solar, wind and wave energy.

Though the design of marine diesel engines are improved day by day, confined to the vessel condition, the marine engines' efficiency is lower than big land equipments, the comprehensive heat efficiency is much lower. Thus, to stop the marine generator and connect shore power can increase the whole energy efficiency and reduce GHG emission. In addition, the land power plant is not much restricted by the space, so the power is easier to get from wind, solar, water and nuclear. When this is the case, the emission reduction would be more obvious.

5 CONCLUSIONS

According to the field data of China international shipping and the calculation of the EEOI baseline, the following conclusions can be summarized:

1. The EEOI baseline of crude oil tanker is fairly low similar with that of bulk carrier. The baselines of container carrier and general cargo vessel are obviously higher than other types of ship.
2. Ship loading ratio, sailing speed and the voyage course have great impact on EEOI. EEOI decreases with the increase of tonnage or loading ratio, and increases with the sailing speed or the complexity of the voyage course.
3. EEOI of Ships with ample cargos and fixed course, such as crude oil tanker and bulk carrier, shows better concentricity and matching degree.

REFERENCES

Yang Su. 2005. The former and existing status of Kyoto Protocol. *Ecological-economic.*

IMO. 2005. INTERIM GUIDELINES FOR VOLUNTARY SHIP CO_2 EMISSION INDEXING FOR USE IN TRIALS. *MEPC/Circ.471.*

IMO. 2009. GUIDELINES FOR VOLUNTARY USE OF THE SHIP ENERGY EFFICIENCY OPERATIONAL INDICATOR (EEOI). *MEPC.1/Circ.684.*

MARINTEK, ECON, DNV. 2000. Study of greenhouse gas emissions from ships. *IMO Issue no.2-31.*

Farrell Alex, Glick Mark. 2000. Natural Gas as a Marine Propulsion Fuel. *Transportation research record. volume number: 1738.*

29. Ecological Risk from Ballast Waters for the Harbour in Świnoujście

Z. Jóźwiak
Maritime University of Szczecin, Institute of Transport Engineering, Poland

ABSTRACT: The purpose of the paper has been pointing out the ballast waters donor ports which for the Świnoujście harbour are of the highest risk category as far as transmitting living alien species is concerned. The species when encountering similar environmental conditions become risk to the local species as food competitors treating the local species as food and expanding in an invasive way and spreading formerly unknown diseases to the local environment.
Moreover, they become dangerous to human health as well as hinder the economy development by reducing fish resources, growing on hydro-technical constructions etc.

1 INTRODUCTION

Water used for vessel ballasting dumped in the port of loading appears to be dangerous for coastal ecosystems. Alien species introductions caused by ballast waters exchange may result in excessive development of the organisms in the new environment and become risk to the local fauna and flora as well as they limit the diversity of living organisms in the coastal basins and river estuaries susceptible for alien species introductions (Doblin and other 2001). In the paper the risk assessment of port waters and coastal ecosystems pollution due to ballast waters dumped into the Świnoujście harbour has been undertaken.

The probability of alien species survival in the new environment is basically affected by the similarity of climate and salinity of waters the alien species originate from as well as the waters of their introduction (Drake L, Meyer A, Forsberg R, and other, 2005). Other significant factors appear to be the duration of the voyage and its characteristic (Santagata S., and other 2008). The more similarities and the shorter the voyage, the more probable it seems for the organism to survive and adjust to the new environment to dominate it as an invasive organism (Polish Harbors Information and Control System – PHICS, UM Szczecin, 2008)..

In the above natural water environment risk assessments of the Świnoujście harbour there have been considered the following risk indicators: water salinity, temperature, time of the voyage duration, the type of the ballast waters donor port (the Baltic or outside the Baltic Sea.

The risk assessment of the alien species introductions has been based on the method of regional risk assessment of alien species introduction for the Baltic Sea (Gollasch S., Leppakoski E.,1999, 2007). Because of the problems with data access (there are no proper databases) there has been applied a model which allows for numerous simplifications but at the same time the assumption that they shall not influence the risk assessment has been made. In the model the following quality factors affecting the ballast waters biological characteristic have been assumed:

- -salinity gradient of the water basin the ballast waters originate from;
- -temperature/climate conditions of the ballast waters donor area;
- -the route of the voyage (the Baltic Sea or outside the Baltic Sea).

The risk assessment of the alien species introductions into the Świnoujście harbour have been brought to comparing similarities concerning environmental conditions of the donor port the ballast waters originate from, and the recipient port, that is the Świnoujście harbour where the ballast waters get dumped, as well as defining the time and the area of the voyage.

The aim of the conducted analysis was to identify the ballast waters donor ports of the highest risk category for the Świnoujście harbour. In order to carry out the analysis it was indispensable to enclose information about the donor ports of the ballast waters dumped in the Świnoujście harbour The list of the donor ports needed to be completed with data concerning the donor ports waters salinity and temperature. Then the time of the vessel journey from the donor port to the Świnoujście harbour had to be defined and the donor ports required being qualified as the Baltic or outside the Baltic Sea ports.

Although for most vessels the place and time of ballasting are recoded, in a proper log book, according to the recommended IMO guidelines, Res. A.868(20), there has not existed any system collecting the data. That is why neither in the ports of the vessels' call, nor in the harbor board such data appear to be accessible.

2 THE RESEARCH METHODOLOGY

In order to define the origin of ballast waters dumped to the water basin of the Świnoujście harbour the database contained in *the Polish Harbors Information and Control System – PHICS* has been made used of (Polish Harbors Information and Control System – PHICS, UM Szczecin, 2008).

On the basis of the data concerning the year 2007 there have been selected all vessels that arrived at the Świnoujście harbour under ballast assuming that their last port of call was the ballast waters donor port. All water ballast donor ports have been assigned to the bio-geographical regions according to the division of „*Large marine ecosystems of the world*" (LMG), according to the guidelines of the IMO Committee of the Sea Environmental Protection contained in the MEPC 162(56) Resolution „*Guidelines for risk assessment under regulation A-4 (G7)*" (Large Marune Ecosystems, Information Portal *http://www.edc.uri.edu/lme,* 2008). Then each of the donor ports' conditions has been compared to the Świnoujście harbour with reference to the water salinity and temperature. There has been calculated the time between the vessel's setting out on a voyage to Świnoujście (taking the ballast waters) and her time of arrival in Świnoujście (ballast waters dump) as well as the donor ports have been located- within the Baltic area (+) and outside the Baltic area (-). It has also been assumed that vessels dumped their ballast waters right after their arrival in Świnoujście. The time of the voyage has been calculated by means of a voyage calculator placed on *World Shipping Register - Sea Distances and Voyage Calculator* (World Shipping Register, 2008). For calculating the voyage time 16 knots has been accepted as the vessel's average speed (Walk M., Modrzejewska H., 2007).

2.1 *Salinity risk assessment*

The risk of the water basin salinity level of the donor port where the ships under ballast arrive from can be high, medium or low (Behrens H.L., LeppäkoskI E., Olenin S. and other, 2005). The risk can be expressed in numbers from 3 to 1. The salinity ranges attributed to each of the particular risk levels for the port of Świnoujście have been presented in Table 1.

Table 1. Port waters salinity risk assessment

Salinity level in the Świnoujście harbour- 1,6‰		
Salinity level [‰]	Risk	Scale of risk
0 - 3	High	3
>3 <7	Medium	2
>7	Low	1

2.2 *Temperature risk assessment*

The temperature risk of donor port waters can be high (3 points), medium (2 points) or low (1 point) depending upon the temperature conditions similarities. According to the areas of ballasting the ships sailing to the Świnoujście harbour there have been outlined 3 risk areas:

1 Eastern-Atlantic-Boreal Region EAB – high risk zone –3 points.
2 Mediterranean-Atlantic Region MA – medium risk zone- 2 points.
3 Western-Atlantic-Boreal Region WAB – low risk zone – 1 point

2.3 *Voyage time risk assessment*

The ballast water tests have proved that when the voyage time is getting prolonged the number of the organisms living in the ballast waters decreases.

Thus, short voyages from not distant ports appear to be the highest category risk. Moreover, considerable changes in ballast waters biological composition have been noticed after 3 and 10 days of ballast waters transport in tanks; after the first 3 days the biggest decrease in number of living organisms has occurred; but after 10 days of the journey most of the other left organisms have died (Dicman, 1999)

Risk range related to the voyage time has been presented in Table 3.

Table 3. Voyage time risk

Voyage time [days]	Risk	Scale of risk
<3	High	3
3-10	Medium	2
>10	Low	1

2.4 *Risk assessment of the voyage route*

In order to assess the risk there have been two types of voyages enumerated:

- from the Baltic ports
- from the ports outside the Baltic Sea

In case of voyages in the area of the Baltic Sea the risk concerning the voyage route has been assumed to be low (1point) and high (3 points) in case of donor ports outside the Baltic Sea area.

2.5 *Total risk assessment*

In order to assess the total risk (R) all points achieved for the particular risk factors (salinity-S, temperature-T, voyage time-V_t, voyage route-V_r) have been summed up according to the formula 1:

[illegible]

The maximum potential number of points a donor port may achieve is 12. The accepted total risk according to Gollasch (2007) and other authors may appear on 4 levels as very high, high, medium and low (Tab. 4).

Table 4. Total risk

Risk	Scale of risk [points]
Very high	12
High	11
Medium	9 - 10
Low	≤8

3 DESCRIPTION OF THE RESULTS

Risk assessment has been analyzed for 123 donor ports, that is the ports which are left for the Świnoujście harbour by vessels under ballast due to which enlisting ports of very high and high risk category as well as medium and low risk category has become possible. There have been identified 14 ports whose ballast waters dumped into the Świnoujście harbour cause very high risk of alien species introductions. These are European ports situated by the North Sea (4 - British, 4- German, 2-Dutch, 2-Norwegian, 1 – Belgian, 1- French).

The ports of high risk category are also situated by the North Sea (2 – British, 2- Norwegian).

In case of Świnoujście the donor ports of very high risk category make 5% of all the considered ports. Recapitulation

In 2009 the Świnoujście harbour was entered by vessels arriving from the ports situated at the coast of the Baltic Sea (bio-region 23), the Norwegian Sea (23), the North Sea (24), the coasts of Ireland and Great Britain (26), the coasts of Iberian Peninsula from the Atlantic Ocean (29), the Mediterranean Sea (26) and the north-east coast of the USA (7).

Out of the 123 ports the ballast waters are transported from to the Świnoujście harbour there are 14 (11%) donor ports of very high risk category, 4 (3%) ports of high risk category, 70 (57%) ports of medium risk category and 35 (29%) ports of low risk category (Fig. 1).

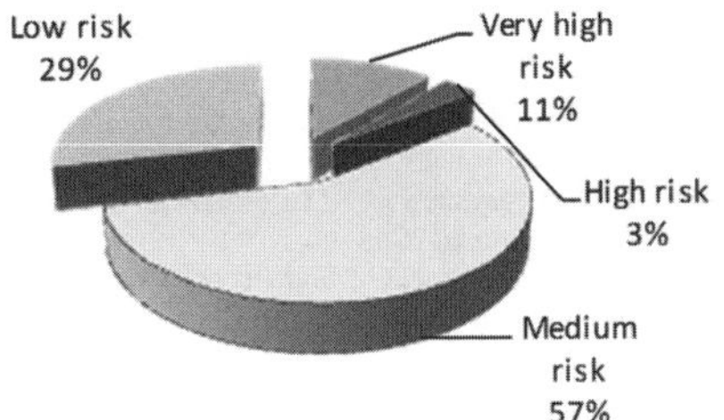

Fig. 1. Donor ports of risk category [%]

Among the ports which the ballast waters taken from appear the most risky to the environment of the Świnoujście harbour there should be enumerated the following ones: Antwerp and Ghent (Belgium), Hamburg, Bremen, Butzfleth and Vierow (Germany), Goole, Keadby, Londonderry and Sutton Bridge (Great Britain), Rotterdam and Terneuzen (Holland), Frederikstad (Norway) and Rochefort (France). It is worth mentioning that these appear to be big ports called at by vessels from all over the world and their waters can be strongly polluted with various kinds of fauna and flora organisms brought there, literally from the whole world.

It seems reasonable to broaden the above research by the other ports of the West Pomerania and expand the research range by testing ballast waters and sediments for the species contained in the transported waters.

REFERENCES

Behrens H.L., Leppäkoski E., Olenin S., Ballast Water Risk Assessment Guidelines for the North Sea and Baltic Sea, Nordic Innovation Centre NT TECHN REPORT 587, Approved 2005-12. Oslo, Norway, 2005. *www.nordicinnovation.net*

Briggs J.C., Marine Zoogeography. McGraw-Hill, New York. 475 pp 1974.

DICKMAN M., ZHANG F., Mid-ocean exchange of container vessel ballast water. 2. Effects of vessel type in the transport of diatoms and dinoflagellates from Manzanillo, Mexico, to Hong Kong, China. *Mar. Ecol.Prog. Ser.* 176: 253-262, 1999.

Drake L, Meyer A, Forsberg R, and other, Potential invasion of microorganisms and pathogens via 'interior hull fouling': biofilms inside ballast water tanks. Biological Invasions 7, 969-982, 2005.

Doblin, M., D. Reid, F. Dobbs, D. and others, Assessment of Transoceanic Nobob Vessels and Low-Salinity Ballast Water as Vectors for Nonindigenous Species Introductions to the Great Lakes. Proceedings of the Second International Conference on Marine Bioinvasions, New Orleans, La., April 9-11, 2001, pp. 34-35, 2001.

Ekman S., Zoogeography of the Sea. Sidgwick & Jackson Ltd., London, 417 pp, 1953.

Gollasch S., Leppakoski E., Initial Risk Assessment of Alien Species in Nordic Coastal waters, Nordic Council of Ministers, Copenhagen, 1999.

Gollasch S., Leppäkoski E., Risk assessment and management scenarios for ballast water mediated species introductions into the Baltic Sea. Aquatic Invasions Volume 2, Issue 4: 313-340, 2007.

Hamer, J., Collin T., Lucas I. Dinoflagellate cysts in ballast tank sediments: Between tank variability. *Mar. Pollut. Bull.* 40: 731-733. 2000.

Large Marine Ecosystems, Information Portal, *http://www.edc.uri.edu/lme,* 2008

Locke, A., Reid D., Leeuwen H.C., and other, 1993. Ballast water exchange as a means of controlling dispersal of freshwater organisms by ships. *Can. J. Fish. Aquat. Sci.* 50: 2086-2093.

Polish Harbors Information and Control System – PHICS, UM Szczecin, 2008.

Santagata S., Zita R. Gasiūnaite Z.R. and other Effect of osmotic shock as a management strategy to reduce transfers of nonindigenous species among low-salinity ports by ships. Aquatic Invasions Volume 3, Issue 1: 61-76, 2008.

Walk M., Modrzejewska H., Ocena ryzyka zawleczenia obcych gatunków na podstawie zaleceń HELCOM – Określenie zagrożenia introdukcji gatunków obcych w Zatoce Gdańskiej na podstawie badan wód balastowych CTO S.A. Gdańsk 2007.

World Shipping Register (*Sea Distances and Voyage Calculator*) - http://www.e-ships.net./dist.htm, 2009

30. A Safety Assurance Assessment Model for an Liquefied Natural Gas (LNG) Tanker Fleet

S. Manivannan
Malaysian Maritime Academy, Malacca, Malaysia

A.K. Ab Saman
Universiti Teknologi Malaysia, Johor, Malaysia

ABSTRACT: With the world's attention on future energy needs focused on LNG; the unprecedented growth phase and globalization era of LNG Shipping activities are inevitable.Along with that the acute shortage of qualified /experienced LNG Sea Officers to manage, operate and maintain the existing and upcoming LNG Tankers is a crucial issue to LNG Shipping Industry as they are the onsite guardian that safeguard and set the standards for the onboard HSSE Assurance. Hence the write-up is an attempt to recommend a recently developed (tailor-made), tested/proven, practical and cost effective solution i.e a Survey Questionnaire based rapid Safety Assurance Assessment Model to safeguard, sustain and further improve any LNG Tanker's Safety Assurance. The Assessment Model was crafted out after an in depth and width study/research into existing and potential future Safety Regimes applicable to an LNG Tanker (including its current Management Approach (Internal Control) – challenges ahead, needs to reform, etc.).The Assessment Model has also taken into consideration the foreseen challenges from the "humans on-site" with reference to already seen and proven historical perspective and impact of human behaviors onto an LNG Tanker's Safety Assurance.

1 INTRODUCTION

With global primary energy demand forecasted to grow at about 1.7%/year from 2002–2030 [1], the world LNG demand is expected to grow at approximately 7%/Year [2].

By 2020 LNG trading via sea set to look much more global [3]. Hence even in current global economic downturn all activities connected with LNG shipping industry are at its Unprecedented Growth Phase [4-7] and is slowly inching into its globalization era [8].

2 STATEMENT OF PROBLEM

In the past LNG shipping industry has leveraged upon its properly trained and experienced LNG Sea Officers to sustain its business success backed by strong HSSE Assurance. Today the utmost concern in LNG shipping industry is the acute shortage of experienced LNG Sea Officers [9] to man the multimillion dollar F1s at sea (LNG Tankers) without jeopardizing its Safety Assurance [10-12].

3 LNG MANPOWER DEMAND

As of 1st April 2009, the global LNG Fleet is forecasted to hit a total of 396 LNG Tankers by year 2012 [13,14]. Hence by 2012, there shall be (at least) a total of about 8700 active/serving LNG Sea Officers to man all the LNG Tankers afloat.

4 LNG MANPOWER ISSUES & ITS POSSIBLE IMPACT ONTO AN LNG TANKER'S SAFETY ASSURANCE.

The above highlighted matters has brought about increased competition and many new challenges to LNG Tankers owners /operators. Following are some Safety Assurance related "concerns" that arise in their attempt to maintain current competitive LNG market position, ventures and commitments:

- Experienced LNG Sea Officers from existing LNG elite group are "poached" using "economic enticements" [15,16]
- "Wrong kind of" unchecked Sea Officers are brought into the industry at higher rank [17].
- LNG Sea Officers might be frequently rotated among various types/class of LNG Tankers [18], leading to LNG Sea Officers/crews (strangers)

cobbled together with little time to develop mutual trust [19].
- Crewing instability can lead to serious deterioration of the relationship between LNG Sea Officers onboard and management ashore within any LNG Tankers operators [20].
- Globally younger generation of Sea Officers ("Y Generation") are withdrawing from the industry prematurely [21].

In conclusion worldwide shortage of LNG experienced Sea Officers can lead to poor decline in Safety Assurance [17,18].

5 CURRENT STATUS AND PROPOSAL

Many LNG Fleet owning /operating companies already feeling the pinch of "concerns" highlighted above. Moving forward, to safeguard, sustain and further improve LNG shipping industry's trademark i.e excellent Safety Assurance track record [22]; **customized, rapid, practical and cost effective solutions are desired.**

However before describing one of such (proposed) solution, let's revisit the typical /existing Safety Assurance regimes of a globally trading LNG Tanker.

6 LNG TANKERS – EXISTING/TYPICAL HSSE REGIMES

6.1 *During Building And At The Point Of Delivery*

Today during construction stage, each LNG Tanker is closely supervised by owner's representatives and appointed Classification Society's surveyors.

These people are entrusted to ensure that a New Building strictly complies (at least) with 17 latest Maritime Rules and Regulations required by Flag State, International Maritime Organization (IMO – Load Line, Tonnage, SOLAS , STCW, ISPS Code, IGC Code, ISM Code, International Convention for Prevention of COLREGs, MARPOL, GMDSS), US Code of Federal Regulations (33 CFR, 46 CFR), US Port and Tanker Safety Act, Suez Canal Authority (SCA), ILO Codes and other Rules & Regulations as decided by owner.

6.2 *In Service*

Upon delivery, during in service for globally trading LNG Shipping Company (hence its LNG Tankers) are expected to complied with Safety Regimes i.e Inspection and Vetting related to or required by ISM, Terminal, SIRE, CDI , Class, Port State Control (PSC) Inspection, Change Of Status, Structural Review, Investigation, Performances and Benchmarking.

7 AN LNG TANKER'S EXISTING HSSE ASSURANCE REGIMES MANAGEMENT

7.1 *Internal Control (IC) ManagementConcept*

Today onboard LNG Tankers almost all the above listed Safety Assurance regimes are managed by its LNG Sea Officers using "Internal Control" (IC) Management Concept which concentrates on the Obligations, Systems, Interfaces and Procedures [23,24]. Generally IC Management Concept has a *"richness"* which is difficult to communicate.

7.2 *IC Management Concept – Challenges Ahead*

The implementation of Safety regimes using IC Management Concept within any industry tends to be "mechanical", with focus on meeting minimal requirements. The approach hence leads to initial improvements in Safety performance that tends to "plateau" after some time [25].

With reference to previously discussed "concerns", LNG Tanker owners/operators need to do more then just "mechanical implementation" of onboard Safety regime.

The implementation shall be elevated to a level where everyone understand, internalize, adapt, adopt, practice, agree and promote on the values of positive Safety behaviors.

7.3 *IC Management Concept – How to Reform?*

To harness "hard to communicate" IC Management Concept richness, its implementation method (model) needs to be fine tuned. The model shall encourage "scientific objectivity" i.e *exposing risk evaluations and decisions to intelligent debate, critics and amendment by people affected by the risk* [26-29].

8 AN LNG TANKER HSSE ASSURANCE

8.1 *Historical Perspective & Future*

Since the beginning of LNG shipping business (in early 1960's), there has been efforts and progress in reducing and keeping the industry's Safety risks to As Low As Reasonably Practicable (ALARP). First generation of LNG ships (about 1960's – 1980's) benefited from its "design" by sustaining its intrinsic "engineering safety". Second generation of LNG ships (about 1980's – 2000's) benefited further through improvement in Safety Management Systems. Today taking note the matters discussed in previous sections; current and future generation (post 2000's) LNG ships' Safety Assurance can only be safeguarded by the integration of and changes to existing organizational culture, personal behavior management and management attitudes [30].

8.2 *Impact Of Onsite Human Behavior*

Ultimately onboard any LNG Tanker its Sea Officers' *"behaviors"* that ensures onsite Safety Assurance and status [22]. Research findings by UK P & I Club and SHELL on *"human behaviour"* [31-33] further elaborate the above statement.

1 The plan that people make in their mind centers around "questions" related to the expected action's –
  - outcome,
  - perceived gap (present Vs ideal) and
  - own ability
2 Individuals' reaction to above questions depend on their beliefs, perceptions, management methods and working environment.
3 Making known the Safety Management Systems' key elements/requirements is crucial for its effective implementation.
4 Verifying whether the person "responsible" understands the above key elements/ requirements is important.
5 Personal proactive intervention through the application of *"Hearts and Mind"* is crucial

The research concluded that continuous improvement in Safety Assurance requires a deeper education/ embedding of the Health, Safety, Security & Environment Management Systems (HSSE MS). People shall be motivated to operate the elements of the HSSE MS, because they believe in it ("want to"), rather than that they are being forced ("have to").

8.3 *Driving Force*

From the above it is a fact that an LNG ship /fleet can improve and sustain its Safety Assurance when its LNG Sea Officers' (i.e its driving force) "hearts and minds" are tactfully addressed. With onboard "educated /reminded" LNG Sea Officers and "checked" /known Safety Assurance status, future "hearts and minds" related initiatives (e.g Behavior Based Safety (BBS), etc) can be easily rolled-out and implemented.

9 PROPOSAL – A RAPID CUSTOMIZED HSSE ASSURANCE ASSESSMENT MODEL FOR AN LNG TANKER

Taking note all the above discussed matters, ideally for educating/assessing Safety Assurance onboard an LNG Tanker, the focus and scope (of key Safety elements) shall expand /cover beyond the typical existing HSSE Regimes.

The above can be practically approached via an "one (1) comprehensive" customized Survey Questionnaires i.e a Rapid Safety Assurance Assessment Model.

The following write up further describe the model.

10 DEVELOPMENT OF CUSTOMIZED RAPID HSSE ASSURANCE ASSESSMENT MODEL FOR LNG TANKER

A "one (1) comprehensive" customized Survey Questionnaires (Rapid Assessment Model) was developed adopting *"process approach"*. The following activities were carried :

1 An in depth study of:
  - Latest study/research (e.g reports, papers, articles, statistics, etc) on or related to Safety Assurance management in maritime and various high risk industries.
  - Latest 17 mandatory Regulatory Requirements applicable to globally trading LNG ships.
  - Typical 19 Safety Assurance related Inspections, Vetting and Other Initiatives imposed upon/adopted by globally trading LNG ships.
  - Existing/in use (active) MISC Berhad LNG Fleet's (one of the largest owner /operator of LNG Tankers in the world) Safety Management Systems.
  - MISC Berhad LNG Fleet's Safety Performances and Standards for last two (2) Financial Years (FYs)
  - Nine (9) future (potential) Human Elements and Organizational Factors related to Safety Assurance improvement initiatives that can be adopted by any LNG Fleet.
  - Reflect back 21 years of personal LNG shipping (onboard/on field) and academic experience and exposure.
2 In the process of studying the above (item 1), the elements crucial to ensure *effective* Safety Assurance Regimes and Systems Implementation were critically analyzed and summarized.
3 Resulting from the above (item 2), seven (7) Elements (variables) were identified as "crucial" for effective implementation of Safety Systems/Assurance onboard any LNG Tanker. The seven (7) Elements are:
  - Leadership
  - Policies
  - Resources Management
  - Hazards Management
  - Planning
  - Execution
  - Assurance

The below diagram illustrate the interlink between the above seven (7) elements. (See Figure 1 below)

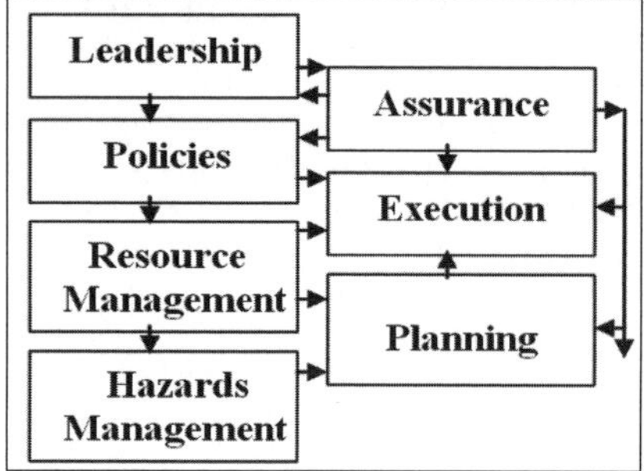

Figure 1. Seven (7) HSSE Assurance Main Elements

To ensure a clear existing status /situational awareness of the research area; a *comparative study* was carried out between the existing typical 36 Safety Regimes for a LNG Tanker against the above seven (7) identified Safety Elements (variables). The comparative study revealed that the existing 36 Safety Regimes address (on average) only 68.6% of the above identified seven (7) Safety Assurance Element.

11 SYNTHESIS OF SURVEY QUESTIONNAIRES

In order to get a fair distribution of data on various Safety Assurance related matters /activities onboard an LNG Tanker all the above detailed seven (7) Element (variables). were treated equally.

The characteristic features of LNG Shipping Safety related matters, challenges, etc. and practicality of conducting an effective Safety Assurance survey (onboard in service/active LNG Tanker) were also well noted during the *synthesis* of the research Survey Questionnaires:

12 CUSTOMIZED RAPID HSSE ASSURANCE

Assessment Model Package Taking note all the above detailed/discussed matters a structured and customized Survey Questionnaires (Rapid Assessment Model) and its "Supportive Documents" were then detailed out under:

- Seven (7) "Main Elements/Topics",
- 37 "Sub Elements/Topics" and
- 252 "Survey Questionnaires".

The below table list down the 37 "Sub Elements/Topics" under the Seven (7) "Main Elements/Topics". (see Table 1 below)

Table 1. Seven (7) Main Elements/Topics and 37 Sub Elements /Topics

1.0	**Leadership**
1.1	Management Visibility
1.2	Proactive Targets Setting
1.3	Informed Involvement
2.0	**Policies**
2.1	Policies Contents & Dissemination
2.2	Strategic Objectives
3.0	**Resources Management**
3.1	Roles, Responsibilities & Accountabilities
3.2	Advisors or Management Representatives
3.3	Resources
3.4	Competency Assurance
3.5	Training
3.6	Contractors / Third Parties
3.7	Communication
3.8	HSSE Committee & Meetings
3.9	Documentation Control
3.10	Checklists & Critical Operation
4.0	**Hazards Management**
4.1	Hazards & Effects Management – General
4.2	Hazards & Effects Identification
4.3	Hazards and Effects Evaluation
4.4	Records of HSSE Hazards and Effects
4.5	Performance Criteria
4.6	Risks Reduction Measures
5.0	**Planning**
5.1	Plans & Initiatives – General
5.2	Critical Facilities & Equipment Integrity
5.3	Procedures & Checklists
5.4	Work/Standing Instructions
5.5	Management Of Change (MOC)
5.6	Emergency Response & Planning
6.0	**Execution**
6.1	Critical Activities & Tasks
6.2	Performance Monitoring
6.3	Records
6.4	Non-Compliance (NCs) & Corrective Actions
6.5	Undesired Events (UDEs) Reporting & Investigation
7.0	**Assurance**
7.1	Assurance Activities
7.2	Assurance Or Audit Plan & Follow-Up
7.3	Internal & External Auditors' Competency
7.4	Contractors/Third Party Auditing
7.5	Management Review

13 RATING SURVEYED ITEMS (PERCENTAGE (%) OF COMPLIANCE) (OR RATING METHOD)

To enable a Survey Respondent to rate i.e to give "opinion on" (points) for a Surveyed Item (Survey Question/Statement); by design for each of the surveyed item either one or both of the following were made available:

1. Compare the "current status" against "minimum requirement"
2. Verify a surveyed item against onboard /onsite "objective evidences".

14 RATING OPTIONS (FIVE (5) POINTS LIKERT MEASUREMENT SCALE)

A dopting the Likert Rating Scale [34] for each surveyed item (Survey technique; Question/Statement) five (5) options were made available for a Survey Respondent. (see Table 2 below)

Table 2. The Surveyed Items - Rating Scale

Point(s)	Survey Respondent's "Opinion"
5	Excellent (E)
4	Good (G)
3	Satisfactory (S)
2	Poor (P)
1	Very Poor (VP)

Hence a Survey Respondent is required to award only one (record his/her feedback or opinion) of the "points".

Using the above tailor-made "measuring instrument", the status of each surveyed items is recorded in a quantitative manner.

15 METHOD OF DATA ANALYSIS

Adopting the above mentioned "Likert Scale"; the tailor-made rapid Safety Assurance Assessment Model was ensured to be compatible with the "Statistical Package For Social Scientist" (SPSS Statistics 17.0) software, leading to various meaningful results on surveyed items can be obtained.

Some of the examples are:

Post Study One (1) (Pre-Treatment/Intervention)

1 Survey Respondents Demographics
2 Standard/Descriptive Statistics (Mean, Standard Deviation and Variance)
3 Distribution Of Feedback (Very Poor, Poor, Satisfactory, Good and Excellent)
4 Normality Of Data Distribution (Skewness and Kurtosis)
5 Reliability (Cronbach's Alpha)
6 Safety Assurance Status Summary (Overall Opinion & Conclusion)
7 Statistical Data Distribution Tests (Kolmogorov-Smirnov (K-S D) & Shapiro- Wilk (S-W) Tests Of Normality)
8 One-Sample T Test (Trial Survey or Pilot Study Vs Actual Study)
9 HSSE Concerns (Written and Interview feedback)
10 HSSE Recommendations (Written and Interview feedback)
11 Survey Findings Reliability (Pearson's Correlation) and Validity (including a Post Survey – Respondents Feedback.

Post Study Two (2), Post-Treatment/Intervention

12 Paired-Sample T Test (Pre-Treatment / Intervention Vs Post-Treatment/Intervention)
13 One-Way Analysis of Variance (One-way ANOVA)

(See Figure 2)

Using the "mean" of Survey Respondents' "opinions" (feedback) for each surveyed Safety Assurance Sub Elements/Topics; its "status" can be determined. Next by calculating out the "average mean" for a group of Safety Assurance Sub Elements /Topics under a Main Element/Topic; the status of a particular Main Element/Topic can be determined. Subsequently with all the 36 Sub Elements/Topics hence the seven (7) Main Elements/Topics "average mean", an entire LNG Fleet's HSSE Assurance status at the point of survey was quantified.

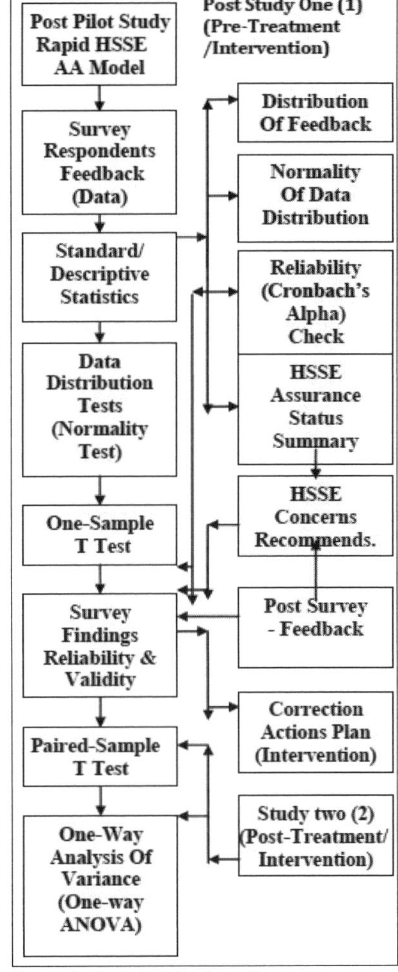

Figure 2. Statistical Analyses/Test

All the above "means" and "average means" can then be directly related to the custmized Safety Assurance Element Assurance – Status Summary and Overall Opinion & Conclusion matrix. This enable better appreciation of the research finding's in term of its "overall opinion" and "conclusion". (see Table 3, below).

Table 3. Safety Element Assurance – Status Summary (Overall Opinion & Conclusion)

Score (%)	Surveyed HSSE Sub Or Main Element – Status Summary
4.0 – 5.0	EXCELLENT (E) Sustain and still scope for continual improvement
3.0 – 3.9	GOOD (G) Sustain and still scope for further ("specific") improvement
2.0 – 2.9	SATISFACTORY (S) Cause for serious concern and scope for "overall" improvement
1.0 – 1.9	POOR (P) Cause for serious concern and immediate enforcement.
0.0 – 0.9	VERY POOR (VP) Cause for serious concern and immediate adoption.

16 IDENTIFYING SHORTCOMINGS (OFIS)

Using the above detailed matrix (Table 3) if a particular Sub or Main Element's/Topic's "means" or "average means" was < **3.000** the particular Sub or Main Element/Topic can be recorded as **"Satisfactory".**

Hence adopting the above described customized method of analyses, shortcomings (OFIs) within any surveyed LNG Tankers'/Fleet's 36 Sub Elements/ Topics hence the seven (7) Main Elements/Topics, crucial for its Safety Assurance can be easily identified/quantified With statistically identified "shortcomings" (OFIs) a structured post survey Improvements/Intervention Plans can be detailed out.

17 COMPREHENSIVE, WELL DISTRIBUTED AND COMPARABLE DATA COLLECTION

To ensure a comprehensive, well distributed and comparable data collection (hence results) from all level of management onboard any surveyed LNG Tanker, the selected portions of the Survey Questionnaires were carefully distributed to relevant pre-identified LNG Sea Officers (by Rank). The approach also ensured that Survey Questionnaires were answered by the rightful Survey Respondents (focal persons).

CASE STUDY – MISC BERHAD LNG TANKER FLEET

18 PILOT STUDY (PRE-TESTING/FINE-TUNING RAPID ASSESSMENT MODEL)

To test out, fine-tune and further improve the Assessment Model prior actual full scale field/onboard survey, a "Pilot Study" was carried onboard three (3) MISC Berhad's LNG Tankers. The Pilot Study statistical results were analyzed using the Statistical Analyses package – *SPSS Statistics 17.0*. Upon completion of the Pilot Study the Rapid Assessment Model was further fine-tuned, improved and finalized.

19 FULL SCALE FIELD/ONBOARD SURVEY FIRST STUDY OR ACTUAL STUDY ONE (1) (PRE-TREATMENT/INTERVENTION)

The finalized Survey Questionnaires (rapid Assessment Model pack) were then sent to ALL 28 MISC Berhad's active/in-service LNG Tankers worldwide.

20 TESTED ASSESSMENT MODEL

A total of 252 active/serving LNG Sea Officers from 28 MISC Berhad's LNG Tankers responded to full scale study.

The Survey Results were analyzed using the latest Statistical Analyses package – *SPSS Statistics 17.0* (as detailed in section 15.0)

The results were then presented to MISC Berhad Top Management. The survey findings were accepted as valid. Relevant "Corrective Actions" were commenced.

After a substantial time lapse (1 year) same survey **(Study Two (2) – (Post-Treatment /Intervention))** were carried out on the same population.

21 CONCLUSION

With reference to LNG shipping industry's foreseen challenges; the way, the existing Safety Regimes onboard LNG Tankers being managed shall be reviewed and tactfully addressed.

It is also important to acknowledge the fact that any proposed recommendations to manage the foreseen "challenges" shall take note of the already seen/proven historical perspective and impact of human behaviors onto an LNG Tanker's Safety Assurance.

Taking note all the above a rapid, practical and cost effective solution to safeguard, sustain and further improve an LNG Tanker's Safety Assurance was crafted adopting a research based "process approach".

The model has been tested by one of the largest owner/operator of LNG Fleet in the world (MISC Berhad) and proven reliable.

REFERENCE

[1] Steve Hull. (2007). Lessons From LNG Trading – Challenges In The Evolution Of An LNG Spot Market. UK: BP.

[2] BP. (2007A). Global Supply Channels. UK: BP.

[3] BP. (2007B). Future (2020) Magnitude And Flow of Pipeline Gas and LNG Trade. UK: BP.

[4] BP. (2007C). Current Magnitude And Flow of Pipeline Gas and LNG Trade. UK: BP.

[5] Chris Lai. (2007). The LNG Carrier Market & An Overview of MISC's LNG Business.Malaysia: MISC Bhd.

[6] Braemar Seascope. (2007). LNG World Shipping Journal (September-October 2007) – LNG Carrier Orders & Deliveries. UK: Riviera Maritime Media Ltd.

[7] W.S. Wayne. (2008). The 58th GPC Meeting & 52nd Panel Meeting Key Note Address. Brunei Darussalam: Society of International Gas Tankers and Terminals Operators (SIGTTO).

[8] Micheal Williams and Richard Harrision. (2006). IQPC Training: LNG Fundamentals. KL : IQPCKL.

[9] Tony Teo. (2007). MMS Singapore Conference 2007 :Impact of Growth in the LNG Sector. US: DNV.

[10] Baltic and International Maritime Council (BIMCO). (2005). World Merchant Shipping Fleet and Manpower Requirements. Denmark: BIMCO.

[11] Mac Hardy, James - Capt. (2006). IQPC LNG Shipping Summit 2006: Key Note Address .Singapore: Society of International Gas Tankers and Terminals Operators (SIGTTO).

[12] UK P & I Club. (2008). Gas Matters – Gas Cargo Claims: How They Are Caused? And How To Avoid Them?: UK. UK P & I Club.

[13] LNG World Shipping Journal (May/June 2008). LNG Carrier Orders and Deliveries, UK : rmm Publications.

[14] Braemar Seascope. (2007). LNG World Shipping Journal (March/April 2009) – LNG Carrier Orders & Deliveries. UK: Riviera Maritime Media Ltd.

[15] SIGTTO. (2005). SIGTTO News. UK: SIGTTO Limited.

[16] William P.Doyle. (2007). Hearing On Safety And Security Of Liquefied Natural Gas And The Impact On Port Operations Broadwater Project Long Island Sound. USA: MEBA.

[17] Keith Bainbridge. (2005). LNG Shipping Solutions: Poaching War for Crews Erupts. UK: Fairplay International Shipping Weekly.

[18] Fairplay. (2005). Near Calamities in Cargo Operations. UK: Fairplay International Shipping Weekly,

[19] Pradeep Chwla. (2006). Officer Crunch Sparks Safety Alarm. Hong Kong: Anglo Eastern Ship Management.

[20] Simon Pressly, (2005). LNG Shipping Solutions: Poaching War for Crews Erupts. UK: Fairplay International Shipping Weekly.

[21] Dr.Stephen Cahoon. (2008). Shipping, Shortages And Generation Y. Singapore. AMC

[22] International Association of Maritime Universities (IAMU). (2005). IAMU Round Table – Maintaining An Unparallel Safety Record In LNG Shipping And Terminals In The Face Of Unprecedented Growth. Korea: IAMU.

[23] Flagstad K.E.(1995). Ph.D Thesis : The Functioning Of Internal Control Reform. Trondheim: Norwegian University of Science and Technology.

[24] Kjellen. (1991). Seminar Paper: Managing Safety in the 2000's - Effects Of Internal Control On Safer Management. USA: Jossy-Bass.

[25] Risktec. (2007A). Behavioral Based Safety Performance Systems. Malaysia: Risktec.

[26] Johnson W.G. (1980), MORT: Safety Assurance Systems. National Safety Council of America : Chicago. NSC Press.

[27] Shrader-Frechette K.S. (1991). Risk and Rationality. Berkely": University of California Press.

[28] Jan Hovden (1984). Models Of Organizations Versus Safety Based On Studies Of The "Internal Control Of SHE" Reform in Norway. Trondheim: Norwegian University of Science and Technology.

[29] Jan Hovden. (1996). Safety Management: Models Of Organizations Versus Safety Management Approaches: A Discussion Based On Studies Of The "Internal Control Of SHE" Reform in Norway : Netherlands. Elsevier Science Ltd.

[30] SIGTTO. (2004). Ship Vetting and Its Application to LNG: The Need to Vet Ships. UK : SIGTTO Limited.

[31] UK P & I Club. (2003). Loss Prevention News – No Room For Error: UK. UK P & I Club.

[32] UK P & I Club. (2004). Loss Prevention News – Accidents – Are They Avoidable: UK. UK P & I Club.

[33] Volkert Zijlker. (2007). The Role of HSE Management System . Historical Perspective and Links With Human Behavior. Netherlands : SHELL E & P.UK P & I Club.

[34] Duane Davis. (2000). Business Research For Decision Making (5th Edition). USA: Duxbury Thomson learning.

Author index